NUREG-1814

# Status of Decommissioning Program

2004 Annual Report

Final Report

U.S. Nuclear Regulatory Commission
Office of Nuclear Material Safety and Safeguards
Washington, DC 20555-0001

# Status of Decommissioning Program

# 2004 Annual Report

# Final Report

Manuscript Completed: January 2005
Date Published: January 2005

Prepared by
J. Buckley

Division of Waste Management and Environmental Protection
Office of Nuclear Material Safety and Safeguards
U.S. Nuclear Regulatory Commission
Washington, DC 20555-0001

## ABSTRACT

This report provides a comprehensive overview of the U.S. Nuclear Regulatory Commission's (NRC's) decommissioning program. Its purpose is to provide a stand-alone reference document which describes the decommissioning process and summarizes the current status of all decommissioning activities including the decommissioning of complex decommissioning sites, commercial reactors, research and test reactors, uranium mill tailings facilities, and fuel cycle facilities. In addition, this report discusses accomplishments in the decommissioning program since publication of the 2003 annual report (SECY-03-0161), and it identifies the key decommissioning program issues which the staff will address in fiscal year (FY) 2005.

## PAPERWORK REDUCTION ACT STATEMENT

The information collections contained in this NUREG are covered by the requirements of Title 10 of the Code of Federal Regulations Parts 19, 20, 30, 33, 34, 35, 36, 39, 40, 51, 70, 72, and 150 which were approved by the Office of Management and Budget (OMB), approval numbers 3150-0044, 0014, 0017, 0015, 0007, 0010, 0158, 0130, 0020, 0021, 0009, 0132, and 0032.

## PUBLIC PROTECTION NOTIFICATION

If a means used to impose an information collection does not display a currently valid OMB control number, NRC may not conduct or sponsor, and a person is not required to respond to, the information collection.

**TABLE OF CONTENTS**

**APPENDICES**

**TABLES**

## ACKNOWLEDGMENTS

This report is a compilation of information from several NRC Offices. As such, many individuals provided valuable assistance in the development and review of this NUREG report.

Participants

Merritt Baker
Kristina Banovac
Helen Chang (Computer Sciences Corporation)
Claudia Craig
Daniel Gillen
John Greeves
Patrick Isaac
Robert Johnson
John Lusher
Robert Nelson
Dominick Orlando
William Ott
Stewart Treby
Cheryl Trottier

## ABBREVIATIONS

| | |
|---|---|
| ACL | alternate concentration limit |
| AEC | U.S. Atomic Energy Commission |
| ASLB | Atomic Safety Licensing Board |
| ATK | Alliant Ordinance and Ground Systems, LLC |
| BTP | branch technical position |
| CERCLA | Comprehensive Environmental Response, Compensation, and Liability Act of 1980 |
| CFR | *Code of Federal Regulations* |
| Ci | curie |
| Co | cobalt |
| CRADAL | Computerized Risk Assessment and Data Analysis Lab |
| Cs | cesium |
| CSM | Consultant Services Meeting |
| CY | calendar year |
| DCGL | Derived Concentration Guideline Level |
| DGSNR | Directorate General for Nuclear Safety Agency |
| DOE | U.S. Department of Energy |
| DP | Decommissioning Plan |
| DU | depleted uranium |
| DWMEP | Division of Waste Management and Environmental Protection |
| EA | environmental assessment |
| EIS | environmental impact statement |
| EPA | U.S. Environmental Protection Agency |
| FA | financial assurance |
| FCSS | Division of Fuel Cycle Safety and Safeguards |
| FONSI | finding of no significant impact |
| FSS | final status survey |
| FSSP | final status survey plan |

| | |
|---|---|
| FSSR | final status survey report |
| FTE | full-time equivalents |
| FUSRAP | formerly utilized sites remedial action program |
| FY | fiscal year |
| g | gram |
| GSA | U.S. General Services Administration |
| IC | institutional control |
| IAEA | International Atomic Energy Agency |
| ISCMEM | Interagency Steering Committee on Multimedia Environmental Models |
| ISCORS | Interagency Steering Committee on Radiation Standards |
| ISFSI | independent spent fuel storage installation |
| ISL | in situ leach |
| km | kilometers |
| kW | kilowatt |
| L | liter |
| LLW | low-level waste |
| LTP | License Termination Plan |
| LTR | License Termination Rule |
| LTSP | Long Term Surveillance Plan |
| MARLAP | Multi-Agency Radiation Laboratory Analytical Protocols |
| MARSSIM | Multi-Agency Radiation Survey and Site Investigation Manual |
| MDA | minimum detectable activity |
| MDEQ | Michigan Department of Environmental Quality |
| MED-AEC | Manhattan Engineering District and the Atomic Energy Commission |
| MOU | memorandum of understanding |
| mrem | millirem |
| NEA | Nuclear Energy Agency |
| NMSS | Office of Nuclear Material Safety and Safeguards |
| NPL | national priorities list |

| | |
|---|---|
| NRC | U.S. Nuclear Regulatory Commission |
| NRR | Office of Nuclear Reactor Regulation |
| ODEQ | Oklahoma Department of Environmental Quality |
| OECD | Organization for Economic Cooperation and Development |
| OGC | Office of the General Counsel |
| OIP | Office of International Programs |
| OMB | Office of Management and Budget |
| ORISE | Oak Ridge Institute for Science and Education |
| ORNL | Oak Ridge National Laboratory |
| PA | preliminary assessment |
| PADEP | Pennsylvania Department of Environmental Protection |
| PART | Program Assessment Rating Tool |
| POTW | Publicly-Owned Treatment Works |
| PSDAR | Post-Shutdown Decommissioning Activities Report |
| Pu | plutonium |
| R&D | research and development |
| RAI | request for additional information |
| RES | Office of Research |
| RIS | Regulatory Issue Summary |
| ROD | Record of Decision |
| RP | Remediation Plan |
| SDMP | Site Decommissioning Management Plan |
| SER | safety evaluation report |
| SNM | special nuclear material |
| Sr | strontium |
| SRM | staff requirements memorandum |
| SRP | Standard Review Plan |
| STP | Office of State and Tribal Programs |
| TBD | to be determined |

| | |
|---|---|
| Tc | technetium |
| Th | thorium |
| TS | technical specification |
| U | uranium |
| UMTRCA | Uranium Mill Tailings Radiation Control Act |
| USACE | U.S. Army Corps of Engineers |
| WASSC | Waste Safety Standards Committee |
| WCS | Waste Control Specialists |
| WDEQ | Wyoming Department of Environmental Quality |
| WPDD | Working Party on Decommissioning and Dismantlement |
| WVDP | West Valley Demonstration Project |
| yr | year |

# ALPHABETICAL LISTING OF SITE SUMMARIES BY SITE CATEGORY

## SITE SUMMARIES FOR DECOMMISSIONING POWER REACTORS

## SITE SUMMARIES FOR RESEARCH AND TEST REACTORS

# SITE SUMMARIES FOR CURRENT COMPLEX DECOMMISSIONING SITES

## SITE SUMMARIES FOR DECOMMISSIONING TITLE II SITES

## SITE SUMMARIES FOR DECOMMISSIONING FUEL CYCLE FACILITIES

1.    **Introduction**

This report provides a comprehensive status of the U.S. Nuclear Regulatory Commission's (NRC's) decommissioning program. Its purpose is to provide a stand-alone reference document that describes the decommissioning process and summarizes the status of all decommissioning activities since the last report, through August 1, 2004, including the decommissioning of commercial reactors, research and test reactors, complex sites, uranium mill tailings facilities, and fuel cycle facilities. In addition, this report discusses accomplishments in the decommissioning program since last year's report (SECY-03-0161), and it identifies key decommissioning program issues which the staff will address in the coming year.

2.    **Decommissioning Sites**

NRC regulates the decontamination and decommissioning of materials and fuel cycle facilities, power reactors, research and test reactors, and uranium recovery facilities, with the ultimate goal of license termination. A broad spectrum of activities associated with these program functions is discussed in this report.

On June 17, 2004, the elimination of the Site Decommissioning Management Plan (SDMP) designation was announced in the *Federal Register* (69 *Federal Register* 33946). Instead, NRC will manage all materials decommissioning sites as "complex sites" under a comprehensive decommissioning program. The SDMP designation will be used in this paper only to describe decommissioning activities which have taken place prior to June 17, 2004.

Approximately 300 materials licenses are terminated each year. Most of these license terminations are routine, and the sites require little, if any, remediation to meet NRC's unrestricted release criteria. The decommissioning program includes termination of licenses that are not routine because the sites involve more complex decommissioning activities.

Currently, there are 20 nuclear power reactors, 17 research and test reactors, 43 complex decommissioning materials facilities, 3 fuel cycle facilities, and 14 uranium recovery facilities that are undergoing non-routine decommissioning or are in long-term safe storage, under NRC jurisdiction. Table 2–1 provides a listing of all sites undergoing decommissioning, by State.

Through the Agreement State Program, thirty-three States have signed formal agreements with NRC, by which those States have assumed regulatory responsibility over certain byproduct, source, and small quantities of special nuclear material, including decommissioning at these sites. Agreement States do not have regulatory authority over operating or decommissioning nuclear power plants.

**Table 2–1**
**Sites Undergoing Decommissioning—By State**

| Name | Location | Facility Type | Date Decomm. Complete | Site Summ. Pg. No. |
|---|---|---|---|---|
| Department of the Army | Fort McClellan, AL | Complex | 6/05 | Page C-10 |
| | | | | |
| General Atomics | San Diego, CA | FC | TBD | Page E-2 |
| General Atomics – TRIGA Mark F | San Diego, CA | R&T-R | TBD | Page B-4 |
| General Atomics – TRIGA Mark I | San Diego, CA | R&T-R | TBD | Page B-5 |
| General Electric Co. – GETR | Sunol, CA | R&T-R | TBD | Page B-6 |
| General Electric Co. – VESR | Alameda, CA | R&T-R | TBD | Page B-7 |
| Humboldt Bay | Eureka, CA | P-R | TBD | Page A-5 |
| Rancho Seco | Sacramento, CA | P-R | 2008 | Page A-12 |
| San Onofre – Unit 1 | San Clemente, CA | P-R | TBD | Page A-13 |
| Vallecitos | Pleasanton, CA | P-R | TBD | Page A-17 |
| | | | | |
| ABB Prospects, Inc. | Windsor, CT | Complex | 12/07 | Page C-3 |
| Haddam Neck – Connecticut Yankee | Meriden, CT | P-R | 2007 | Page A-4 |
| Millstone – Unit 1 | Waterford, CT | P-R | TBD | Page A-9 |
| UNC Naval Products | New Haven, CT | Complex | TBD | Page C-42 |
| | | | | |
| Eglin Air Force Base | Walton County, FL | Complex | 2005 | Page C-12 |
| | | | | |
| Salmon River | Salmon, ID | Complex | TBD | Page C-36 |
| | | | | |
| Dresden – Unit 1 | Dresden, IL | P-R | TBD | Page A-2 |
| Engelhard Minerals – Illinois | Great Lakes, IL | Complex | 12/05 | Page C-13 |
| Honeywell | Metropolis, IL | FC | TBD | Page E-3 |
| University of Illinois | Urbana, IL | R&T-R | TBD | Page B-12 |
| Zion – Units 1 & 2 | Waukegan, IL | P-R | TBD | Page A-19 |

**Table 2–1**
**Sites Undergoing Decommissioning—By State (continued)**

| Name | Location | Facility Type | Date Decomm. Complete | Site Summ. Pg. No. |
|---|---|---|---|---|
| | | | | |
| Jefferson Proving Ground (Department of Army) | Madison, IN | Complex | TBD | Page C-19 |
| | | | | |
| Maine Yankee | Wiscasset, ME | P-R | 6/05 | Page A-8 |
| | | | | |
| Yankee Rowe | Greenfield, MA | P-R | 2021 | Page A-18 |
| | | | | |
| AAR Manufacturing, Inc. | Livonia, MI | Complex | 1/07 | Page C-1 |
| Big Rock Point | Charlevoix, MI | P-R | 2012 | Page A-1 |
| Dow Chemical Company | Bay City, MI | Complex | 4/06 | Page C-11 |
| Fermi – Unit 1 | Newport, MI | P-R | 2008 | Page A-3 |
| Ford Nuclear Reactor | Ann Arbor, MI | R&T-R | TBD | Page B-3 |
| Michigan Department of Natural Resources | Kawkawlin, MI | Complex | 10/06 | Page C-27 |
| NWI Breckenridge | Breckenridge, MI | Complex | 12/04 | Page C-29 |
| SCA Services (SCA) | Kawkawlin, MI | Complex | 7/11 | Page C-37 |
| | | | | |
| Alliant Ordinance and Ground Systems (ATK) | Arden Hills, MN | Complex | 12/04 | Page C-4 |
| | | | | |
| Mallinckrodt Chemical, Inc. | St. Louis, MO | Complex | 7/08 | Page C-26 |
| Westinghouse Electric (Hematite Facility) | Jefferson City, MO | Complex | 3/10 | Page C-47 |
| | | | | |
| Veterans Administration | Omaha, NE | R&T-R | TBD | Page B-16 |
| | | | | |
| Heritage Minerals | Lakehurst, NJ | Complex | 6/05 | Page C-17 |
| Shieldalloy Metallurgical Corp. | Newfield, NJ | Complex | 9/10 | Page C-38 |
| Stepan Chemical Company | Maywood, NJ | Complex | 9/09 | Page C-39 |

## Table 2–1
## Sites Undergoing Decommissioning—By State (continued)

| Name | Location | Facility Type | Date Decomm. Complete | Site Summ. Pg. No. |
|------|----------|---------------|-----------------------|--------------------|
| | | | | |
| Kirtland Air Force Base | Albuquerque, NM | Complex | 4/05 | Page C-24 |
| Homestake | Grants, NM | UR | 2015 | Page D-5 |
| Rio Algom – Ambrosia Lake | McKinley Co., NM | UR | 2008 | Page D-9 |
| Sohio L-Bar | Seboyeta, NM | UR | 2004 | Page D-11 |
| United Nuclear Corp. | Church Rock, NM | UR | 2015 | Page D-13 |
| | | | | |
| Cornell University – TRIGA | Ithaca, NY | R&T-R | TBD | Page B-1 |
| Cornell University – ZPR | Ithaca, NY | R&T-R | TBD | Page B-2 |
| Indian Point – Unit 1 | Buchanan, NY | P-R | TBD | Page A-6 |
| Manhattan College | Bronx, NY | R&T-R | TBD | Page B-8 |
| University of Buffalo | Buffalo, NY | R&T-R | TBD | Page B-11 |
| West Valley | West Valley, NY | Complex | TBD | Page C-44 |
| | | | | |
| Battelle Columbus Laboratories | Columbus, OH | Complex | 12/05 | Page C-7 |
| Engelhard Minerals – Ohio | Ravenna, OH | Complex | 3/05 | Page C-14 |
| NASA – Mockup | Sandusky, OH | R&T-R | 2007 | Page B-9 |
| NASA – Plum Brook | Sandusky, OH | R&T-R | 2007 | Page B-10 |
| | | | | |
| FMRI, Inc. (formerly Fansteel) | Muskogee, OK | Complex | 2023+ | Page C-15 |
| Kaiser Aluminum | Tulsa, OK | Complex | 5/07 | Page C-20 |
| Kerr-McGee – Cimarron | Cimarron, OK | Complex | 5/07 | Page C-21 |
| Kerr-McGee – Cushing Refinery Site | Cushing, OK | Complex | 12/05 | Page C-22 |
| Kerr McGee Tech. Center | Oklahoma City, OK | Complex | 12/04 | Page C-23 |
| Sequoyah Fuels Corp. | Gore, OK | UR | 2010 | Page D-10 |

**Table 2–1**
**Sites Undergoing Decommissioning—By State (continued)**

| Name | Location | Facility Type | Date Decomm. Complete | Site Summ. Pg. No. |
|---|---|---|---|---|
| | | | | |
| Trojan | Rainier, OR | P-R | 6/05 | Page A-16 |
| | | | | |
| Babcock & Wilcox (Shallow Land Disposal Area) | Vandergrift, PA | Complex | 10/09 | Page C-6 |
| Cabot Performance Materials, Inc. | Reading, PA | Complex | 9/05 | Page C-8 |
| Curtis-Wright Cheswick | Cheswick, PA | Complex | 12/08 | Page C-9 |
| Kiski Valley Water Pollution Control Authority | Vandergrift, PA | Complex | 11/04 | Page C-25 |
| Molycorp, Inc. – Washington | Wash., PA | Complex | 10/06 | Page C-28 |
| Peach Bottom – Unit 1 | Delta, PA | P-R | 2014 | Page A-11 |
| Quehanna (formerly Permagrain Products, Inc.) | Media, PA | Complex | 12/04 | Page C-31 |
| Royersford Wastewater Treatment Facility | Royersford, PA | Complex | TBD | Page C-32 |
| Safety Light Corp. | Bloomsburg, PA | Complex | TBD | Page C-34 |
| Saxton | Saxton, PA | P-R | 2004 | Page A-14 |
| Superior Steel | Pittsburgh, PA | Complex | TBD | Page C-40 |
| Three Mile Island – Unit 2 | Harrisburg, PA | P-R | TBD | Page A-15 |
| Westinghouse | New Stanton, PA | R&T-R | 2005 | Page B-17 |
| Westinghouse Electric Company | Blairsville, PA | Complex | 12/04 | Page C-46 |
| Westinghouse Electric Company, Waltz Mill | Madison, PA | Complex | 8/05 | Page C-49 |
| Whittaker Corp. | Greenville, PA | Complex | 9/05 | Page C-50 |
| | | | | |
| Augustana College | Sioux Falls, SD | Complex | 9/04 | Page C-5 |
| Pathfinder | Sioux Falls, SD | Complex | 4/06 | Page C-30 |
| | | | | |
| Union Carbide Corp. | Lawrenceburg, TN | Complex | 12/05 | Page C-43 |

**Table 2–1**
**Sites Undergoing Decommissioning—By State (continued)**

| Name | Location | Facility Type | Date Decomm. Complete | Site Summ. Pg. No. |
|---|---|---|---|---|
| | | | | |
| Nuclear Ship Savannah | Newport News, VA | P-R | TBD | Page A-10 |
| University of Virginia | Charlottesville, VA | R&T-R | TBD | Page B-13 |
| University of Virginia – Cavalier | Charlottesville, VA | R&T-R | TBD | Page B-14 |
| | | | | |
| Framatome Richland | Richland, WA | FC | TBD | Page E-1 |
| University of Washington | Seattle, WA | R&T-R | TBD | Page B-15 |
| | | | | |
| Homer Laughlin | Newell, WV | Complex | TBD | Page C-18 |
| | | | | |
| Lacrosse | La Crosse, WI | P-R | TBD | Page A-7 |
| | | | | |
| American Nuclear Corp. | Gas Hills, WY | UR | 2007 | Page D-1 |
| Bear Creek | Converse City, WY | UR | 2004 | Page D-2 |
| COGEMA Mining, Inc. | Johnson & Campbell Counties, WY | UR | 2007 | Page D-3 |
| ExxonMobil Highlands | Converse Co., WY | UR | 2005 | Page D-4 |
| Pathfinder – Lucky MC | Gas Hills, WY | UR | 2005 | Page D-6 |
| Pathfinder – Shirley Basin | Shirley Basin, WY | UR | 2007 | Page D-7 |
| Petrotomics | Shirley Basin, WY | UR | 2004 | Page D-8 |
| Umetco Minerals Corp. | East Gas Hills, WY | UR | 2006 | Page D-12 |
| Western Nuclear, Inc. – Split Rock | Jeffrey City, WY | UR | 2007 | Page D-14 |

NOTE:

• Abbreviations used in this table include Complex for Complex Decommissioning Materials Facility; FC, Fuel Cycle Facility; P-R, Power Reactor Facility; R&T-R, Research and Test Reactor Facility; and UR, Uranium Recovery Facility.

## 2.1    Nuclear Power Reactors

The Office of Nuclear Material Safety and Safeguards (NMSS) currently has regulatory project management responsibility for 15 decommissioning power reactors.  The Office of Nuclear Reactor Regulation (NRR) has project management responsibility for two decommissioning reactors (Indian Point – Unit 1, Millstone – Unit 1) because extensive stakeholder interest in these sites (for both the operating and decommissioning units) makes it more efficient for NRR to perform, as a single point of contact, project management responsibilities for the permanently shutdown units.  In addition, NRR has project management for three decommissioning early demonstration reactors—Vallecitos, Nuclear Ship Savannah, and Saxton.  In Section 9.1.2, Table 9–1, identifies the power reactors undergoing decommissioning.  Plant status summaries for all decommissioning reactors are provided in Appendix A.  The staff currently is reviewing the License Termination Plans (LTPs) for Big Rock Point (submitted in April 2003) and for Yankee Rowe (submitted in November 2003).

## 2.2    Research and Test Reactors

NRR provides project management and inspection oversight for 17 decommissioning research and test reactors.  Currently, 13 research and test reactors have decommissioning orders or amendments.  Additionally, three research and test reactors are in "possession-only" status, either waiting for shutdown of another research or test reactor at the site, or for removal of the fuel from the site by the U.S. Department of Energy (DOE).  One research and test reactor is preparing to submit a decommissioning amendment request.  Further, 3 of the 13 research and test reactors with decommissioning orders or amendments, and 1 of the 3 research and test reactors in possession-only status still have fuel in storage at the reactor.  In Section 9.2.2, Table 9–2 identifies the research and test reactors undergoing decommissioning.  Plant status summaries are provided in Appendix B.

## 2.3    Complex Decommissioning Materials Facilities

On June 17, 2004, the staff published a Notice in the *Federal Register* (69 *Federal Register* 33946) to announce that NRC has decided to eliminate the SDMP designation for sites and manage the SDMP sites as "complex sites" under a comprehensive decommissioning program.  See Section 7 for a more detailed discussion of this action.  Currently, there are 43 complex decommissioning materials facilities (see Section 10.1.1, Table 10–1).  Since last year's status report, five sites were removed from the complex site list:  (1) Babcock & Wilcox – Parks Township; (2) Envirotest Laboratories; (3) Molycorp, Inc. – York; (4) University of Wyoming; and (5) Watertown – GSA.

NRC completed its evaluation of formerly licensed sites under the Oak Ridge National Laboratory (ORNL) Terminated License Review Project in September 2001.  As a result of the ORNL review, and subsequent follow up by the Regions, 42 formerly licensed sites were found to have residual contamination levels exceeding NRC's criteria for unrestricted release.  After successful remediation, 20 of these sites have been closed, and 11 have been closed by transfer to Agreement States or a Federal entity for closure under their oversight programs.  Eleven sites remain open and are managed as complex decommissioning materials facilities.

In calendar year (CY) 2004, the DWMEP staff continued to implement its comprehensive integrated plan for successfully bringing complex decommissioning sites to closure.  Site status summaries are maintained, for each complex decommissioning site, and are provided in Appendix C. These summaries describe the status of each site and identify the current technical and regulatory issues impacting completion of decommissioning. The staff also maintains schedules (Gantt charts) for each site, which are updated quarterly, to guide the management of decommissioning activities. The Gantt charts identify all major decommissioning activities and schedules for completion. For those licensees that have submitted a decommissioning plan (DP), the schedules are based on an assessment of the complexity of the DP review. For those licensees that have not submitted a DP, the schedules are based on other licensee information available and the anticipated decommissioning approach. To date, 6 of the 43 complex decommissioning sites have not yet submitted DPs, and NRC is currently reviewing 10 DPs.

## 2.4    Uranium Recovery Facilities

NMSS provides project management and technical review for decommissioning and reclamation of facilities that are regulated under 10 CFR Part 40, Appendix A. These licensees include conventional uranium mills and in situ leach (ISL) facilities. Currently, there are 14 NRC-licensed [Uranium Mill Tailings Radiation Control Act (UMTRCA) Title II] sites in decommissioning. NRC recently approved a request from Utah to amend its Agreement under Section 274 of the Atomic Energy Act to assume regulatory authority over certain additional radioactive material within the State. Effective August 16, 2004, NRC transferred to Utah the responsibility for licensing, inspection, enforcement, and rulemaking activities for uranium and thorium milling operations and mill tailings and other wastes, known as 11e.(2) byproduct material. Two decommissioning UMTRCA Title II sites were thus transferred to Utah: Plateau Resources – Shootaring Canyon and Rio Algom – Lisbon. In Section 10.2.2, Table 10–2 identifies the Title II decommissioning sites. Site status summaries are provided in Appendix D.

## 2.5    Fuel Cycle Facilities

NMSS provides licensing oversight and decommissioning project management to fuel cycle facilities, including conversion plants, enrichment plants, and fuel manufacturing plants. Most of these facilities have been in operation for 20 or more years. As technology improves and operations at these facilities change, there are often unused areas on the site with residual contamination. The NRC staff continues to work closely with the States and EPA to regulate remediation of unused portions of fuel cycle facilities. In 2004, one conversion facility (Honeywell) and two fuel manufacturers (Framatome Richland and General Atomics) continued some decommissioning activities, although all are still operating. In Section 10.3.2, Table 10–3 identifies the fuel cycle facilities undergoing decommissioning. Facility status summaries are provided in Appendix E.

## 3.    Guidance and Rulemaking Activities

In previous years, the staff considered broad-scope regulatory improvements for decommissioning nuclear power plants in the areas of security, emergency planning, and insurance. However, because of continuing efforts by the staff to reassess vulnerabilities and redefine the threats in the area of safeguards and security, the priority for decommissioning regulatory improvements for decommissioning reactors has been reduced. A relatively small number of nuclear power plants are undergoing decommissioning, and the staff does not anticipate additional nuclear power plants decommissioning in the near future. Given the absence of any additional, anticipatable nuclear power plant decommissionings and the uncertainties related to safeguards and security regulation, resources are being deferred for future nuclear power plant decommissioning rulemakings that are currently in progress or related to security matters. Resources for nuclear-power-plant decommissioning rulemakings that are not currently in progress or related to security matters will not be included in the FY 2005 or FY 2006 budgets. If any plants do unexpectedly shut down permanently, decommissioning regulatory issues would continue to be addressed through the amendment and exemption process in a manner similar to the current practice.

In an SRM dated June 6, 2001, the Commission directed the staff to develop a rulemaking to amend the financial assurance requirements for materials licensees in 10 CFR Parts 30, 40, and 70. The staff had notified the Commission of its intent to amend the financial assurance requirements in SECY-01-0084, "Rulemaking Plan: Financial Assurance Amendments for Materials Licensees." The proposed rule was published in the *Federal Register* on October 7, 2002, and the comment period closed on December 23, 2002. The final rule was published in the *Federal Register* on October 3, 2003. The following changes were included in the final rule: (a) large sealed source licensees (such as irradiation facilities), whose authorized possession limits exceed specified amounts, and all waste brokers must provide financial assurance based on site-specific decommissioning cost estimates rather than using an amount prescribed by regulation; (b) the prescribed amounts were increased by 50 percent; and (c) licensees using a decommissioning cost estimate have to update it at least every 3 years.

In SECY-03-0069 (Results of the License Termination Rule Analysis), the staff recommended, in part, that the Commission approve a new rulemaking to reduce the potential for future legacy sites by adding and revising requirements for financial assurance and licensee monitoring, reporting, and remediation. In an SRM dated November 17, 2003, the Commission approved the rulemaking. In April 2004, the staff recommended that it bypass the rulemaking plan stage and proceed with the proposed rule. The Commission also approved that recommendation. The current schedule requires that the staff complete the proposed rule in FY 2006.

On October 25, 2002, the Commission directed the staff to conduct an enhanced participatory rulemaking on the disposition of solid materials. Currently, the staff is considering comments received from stakeholders in letters and e-mails and collected at the public workshop held on May 21–22, 2003; reviewing relevant standards and documents; and maintaining awareness of efforts being conducted by State, Federal, and international organizations. The staff is preparing a generic environmental impact statement (EIS) that evaluates impacts and costs of various alternatives and developing guidance for implementing the rulemaking. The current schedule is to send a rulemaking package to the Commission in March 2005 for issuance as a proposed rule for public comment.

Changes to the decommissioning guidance associated with the LTR Analysis issues are planned for FY 2005 and FY 2006. Section 7 provides details on these planned changes.

4.    **Research Activities**

The Office of Research (RES) continued to provide information to NMSS to support assessments of public exposure to environmental releases of radioactive material from site decommissioning. Several examples of the types of research information provided are discussed in the following paragraphs.

RES staff has several projects underway to improve existing dose modeling codes. Enhancements to the RESRAD–OFFSITE code for modeling the potential impact due to offsite releases was provided to licensing staff for testing, and an updated RESRAD–BIOTA code that will assess impacts to biota was provided for review and testing. A training course on the HYDRUS code that can be used to model groundwater transport was provided to staff. Training was also provided on a beta version of the FRAMES2 software that will be useful to the licensing staff in conducting assessments of environmental system performance at complex sites. The beta version was also provided to the staff for participation in testing this version of the software. The FRAMES2 platform is the current choice for a modularized system that will be able to include codes like HYDRUS in developing a site specific model. Each of these codes represents an increase in the capability of the codes available to provide realistic analyses of complex decommissioning sites.

In addition, RES made a significant contribution to the rulemaking effort on controlling the release of solid materials by (a) completing the assessment of collective doses for potential release strategies; (b) assessing the potential doses from reuse of released soil; (c) developing information for the analysis of conditional use of materials; (d) coordinating the review of draft International Atomic Energy Agency (IAEA) Safety Guide 161 (DS-161) "Application of the Concepts of Exclusion, Exemption and Clearance" and supporting documentation; and (e) completing draft NUREG-1761, "Radiological Surveys for Controlling Release of Solid Materials," and addressing public comments. The majority of these efforts contributed to the technical basis for the EIS for the rulemaking on controlling the release of solid materials.

During the past year, RES staff also continued to support numerous interagency activities. Examples include the development of two manuals, the Multi-Agency Radiation Survey and Site Investigation Manual (MARSSIM) and Multi-Agency Radiation Laboratory Analytical Protocols (MARLAP). Each of these manuals will establish a common approach among Federal agencies for radiological measurements and surveys. RES staff also continued participation in activities of the Interagency Steering Committee on Radiation Standards (ISCORS), including the subcommittees on Recycle and Sewage Sludge, and participation in working groups of the Interagency Steering Committee on Multimedia Environmental Models (ISCMEM). Examples of this work include the following: (1) NUREG-1775, "Interagency Steering Committee on Radiation Standards Assessment of Radioactivity in Sewage Sludge: Survey Results and Analysis," was developed providing results of a national survey of sewage sludge and ash samples obtained from 313 publicly-owned treatment works (POTWs); and (2) Research Information Letter 0303, "Disposal of Radioactive Material into Sanitary Sewer Systems," was issued to inform the user office of research results to evaluate the potential for radioactive materials to concentrate in sanitary sewer systems and the significance of such concentrations.

Major RES activities in 2004-2005 include (a) completion of the technical basis work to support the EIS for the rulemaking on the control of solid materials, (b) continuation of work to increase the realism of dose models used in performance assessment, (c) linkage of FRAMES2 environmental modeling platform to the U.S. Army Corps of Engineers (USACE) ground-water modeling system, (d) extension of uncertainty calculations to include scenario uncertainty, and (e) cooperation with the DOE on development of RESRAD–OFFSITE and RESRAD–BIOTA.

## 5.     International Activities

DWMEP interacts with international organizations and governments in a number of ways including (a) participating in the IAEA, (b) participating in the Organization for Economic Cooperation and Development's Nuclear Energy Agency (OECD/NEA), (c) participating in bilateral and trilateral exchanges with other countries, (d) hosting foreign assignees and providing reciprocal assignments, (e) developing and providing workshops to requesting countries, and (f) providing technical support as needed to the NRC Office of International Programs (OIP).  The NRC  is generally recognized in the international nuclear community as an experienced leader in the decommissioning of nuclear sites.  NRC staff interactions with international organizations and governments allows the NRC  to provide less experienced organizations with insights into decommissioning approaches that are successful, safe  and cost-effective.  It also allows the NRC staff to provide input into the various international guidance and requirements that NRC  and  NRC licensees will need to consider as they interact in a global environment.  A summary of each of these activities is provided below.

### 5.1     International Atomic Energy Agency Activities

The NRC decommissioning staff participated in the development of the IAEA Safety Standards Series.  Within the past year DWMEP supported the IAEA by participating in:

- The Joint Convention on the Safety of Spent Fuel Management and on the Safety of Radioactive Waste Management.  Staff activities included (a) preparation of the U.S. National Report, (b) review and comment on decommissioning and other waste management programs in other Contracting Party Member States, and (c) participation in, and review of, formal presentation of Contracting Party Waste Safety Programs.

- The development of DS-333, "Safety Requirements for the Decommissioning of Nuclear Facilities," Consultant Services Meeting (CSM) held in Vienna, Austria, January 2004.

- The development of DS-332, "Safety Guide on the Removal of Sites and Buildings from Regulatory Control upon the Termination of Practices," CSM held in Vienna, Austria, July 2004.

- The development of the IAEA Action Plan on Decommissioning Nuclear Facilities, for consideration by the IAEA Board of Governors June 2004.

- The development of RS-G-1.7 which addresses the release of materials with low levels of radioactivity (for volumetrically controlled materials).

### 5.2     Organization for Economic Cooperation and Development's Nuclear Energy Agency Activities

The OECD/NEA Radioactive Waste Management Committee serves as the main forum for discussion of topics important to waste management, including decommissioning.  Specific focus is placed on decommissioning by the OECD/NEA Working Party on Decommissioning and Dismantlement (WPDD).  From August 29, 2003, to September 6, 2003, DWMEP staff participated in a NEA WPDD International Seminar on "Strategy Selection for the Decommissioning of Nuclear Facilities."  This seminar was held in conjunction with the regular yearly meeting of the WPDD.  The objective of the seminar was to have focused discussions between implementers (i.e., licensees), regulators, and local communities on the key factors that influence the selection of a particular decommissioning strategy.

In addition to the International Seminar the staff has provided input to NEA on National Fact Sheets, Site Release questionnaires, and National Map and worked with WPDD members to develop a *Decommissioning Safety Case*.

DWMEP staff also provided input to the WPDD Core Group to develop the format of and topics for the Decommissioning Workshop in Rome, Italy, on September 6–10, 2004, in which DWMEP and other NMSS staff participated.

5.3     Bilateral and Trilateral Exchanges with Other Countries

Delegations from France, Hungary, Japan, and Taiwan visited NRC in FY 2004 to discuss many topics associated with radioactive waste management.  Facility decommissioning is usually of keen interest to the visiting delegations.

In addition to hosting individual delegations, the staff participates in a bilateral exchange with the French Directorate General for Nuclear Safety Agency (DGSNR) and a trilateral exchange with Mexico and Canada.  Decommissioning is one of the many topics discussed during the exchanges.  The bilateral exchange with the French takes place twice a year; once in the United States and once in France.  Management and staff from NMSS visited France in October 2003, and the French delegation visited the United States in April 2004.  Each of these visits included site visits to licensed facilities.

The trilateral with Mexico and Canada took place in Ottawa, Canada in July 2004.  Topics included decommissioning experience and regulatory developments.

5.3.1   Hosting Foreign Assignees and Providing Reciprocal Assignments

Through the OIP, an assignee from the People's Republic of China was selected and has begun a 6-month assignment in the Decommissioning Directorate.

5.3.2   Developing and Providing Workshops to Requesting Countries

In March 2004, DWMEP staff conducted a decommissioning workshop in Taipei, Taiwan. Similar workshops are being planned for Russia and Scotland in September 2005.

5.3.3   Providing Technical Support as Needed to the Office of International Programs

DWMEP staff provide technical support to OIP on various decommissioning topics such as:

- Supporting annual Commission briefings on international activities,
- Providing technical input into export/import proposals and requests,
- Responding to international questionnaires and inquiries,
- Supporting meetings of the international council, and
- Reviewing management directives on international interactions and activities.

## 6.    Program Integration

NMSS, NRR, RES, and Regions I, III, and IV share responsibility for decommissioning program activities.  NRR has project management responsibility for all stages of research reactor and test reactor decommissioning and oversight of the initial stages of power reactor decommissioning.  NMSS regulates the decommissioning of nuclear material facilities, fuel cycle facilities, and uranium recovery facilities, and has oversight of power reactors (once the plant has completed regulatory and safety milestones that ensure that the plant more closely represents a materials facility temporarily storing and processing radioactive waste than a commercial power reactor).  The Regions have the lead in the inspection of decommissioning activities and provide project management for several complex materials sites.  RES provides technical support when needed and improved analytical tools to evaluate complex situations involving site contamination.

In a new initiative this year, RES and NMSS management have agreed to establish interoffice working groups to support the decommissioning performance assessments for complex sites such as West Valley.  For that particular site, the working group includes three RES staff with special expertise in performance assessment modeling of natural systems and engineered barriers and extensive knowledge of the West Valley Site.  The major goals of this increased cooperation are (a) to make more realistic licensing decisions and (b) to focus research activities in areas of greatest benefit in addressing licensing issues and on problems revealed through application of new analytical techniques.

RES continued to provide support for the decommissioning program both through the products of its research and the participation of its staff in analyzing difficult issues in performance assessment and MARSSIM implementation.  The research program provides important information in the form of models, supporting information, and training to support the application of research results and models.  RES staff continued to participate in interagency efforts such as ISCORS and the ISCMEM.

The staff continues to take steps to ensure integration of decommissioning activities.  First, NMSS and RES mutually track and manage decommissioning activities.  Second, the Decommissioning Management Board (hereafter, the Board) meets monthly to provide management input on decommissioning activities and issues.  The Board, composed of managers from NMSS, RES, NRR, and the Regions, along with the Office of the General Counsel (OGC), serves as an effective mechanism for integrating inter-Office and inter-Regional program activities and issue resolution.  The Board is a mechanism by which the staff has enhanced intra-agency communication.  In addition, it ensures that NRC's regulatory processes are integrated.

The decommissioning process is becoming more efficient as the staff continues (a) assuming a more proactive role in interacting with licensees undergoing decommissioning, including conducting pre-submittal meetings with licensees; (b) using an expanded acceptance review process, to include a limited technical review, to reduce the need for additional rounds of questions; (c) ensuring that institutional controls and financial assurance requirements are adequate before beginning a technical review of a DP; (d) implementing other procedures (e.g., focused site visits to reduce the number of requests for additional information);

(e) conducting inprocess or side-by-side confirmatory surveys; and (f) relying more heavily on licensees' quality assurance programs rather than conducting large-scale confirmatory surveys.

Furthermore, the staff is incorporating strategies to achieve the performance goals identified as part of the Agency's strategic planning process and Strategic Plans for FYs 2004–2009. Examples of strategies being incorporated include (a) focusing on resolving key issues, such as institutional control for restricted release and partial site release; (b) participating in stakeholder workshops to seek licensee, industry, and public input; (c) updating, consolidating, and risk-informing (i.e., performance-orienting) decommissioning guidance; (d) working with industry to identify and resolve technical and policy issues associated with decommissioning; and continuous refinement of the stakeholder database and Website.

7.      **Issues Raised and Addressed Since the 2003 Annual Report**

7.1     Changes to the Decommissioning Program and Annual Report

In the SRM to SECY 03-0161, "2003 Annual Update—Status of Decommissioning Program," the Commission directed the staff to address (a) expanding the focus and increasing the level of detail in the Annual Decommissioning Report with the intent of developing it as a primary communication document; (b) other changes to the format of the report; and (c) eliminating the SDMP designation.

These following changes are reflected in this report:

*       Site descriptions are provided for all decommissioning sites reported including power reactors, research and test reactors, complex material sites, fuel cycle facilities, and uranium recovery facilities.

*       The site descriptions for all sites discussed in the report use the same format and contain the same level of detail as the sites in the revised 2003 report.

*       The format of the report is revised to address (a) decommissioning programmatic activities since the last report, (b) reactor decommissioning status, (c) complex materials decommissioning status, and (d) decommissioning activities planned for the coming fiscal year.

*       The report provides information on the decommissioning process for materials and reactor facilities.

Because the annual report will contain information that is not expected to change from year to year (i.e., discussion of the materials decommissioning process), the staff will provide the report in the form of a NUREG document every two years beginning with this report. In the odd number years, the staff will publish the report as a shortened paper to the Commission, using references to the decommissioning Website.

To ensure that the Commission and the public will be able to access current information on sites undergoing decommissioning or reclamation activities, site descriptions for all sites undergoing decommissioning will be placed on the NRC Website. These site descriptions will be reviewed on a bi-monthly basis and updated as necessary to ensure that the information is current. Updates will be performed by the respective site Project Managers in NMSS, NRR and the

Regions.  The NUREG report will contain the most current site summaries for each site while the off-year report to the Commission will contain the Website addresses for the sites.

7.2     Elimination of the Site Decommissioning Management Plan and Management of All Sites Undergoing Decommissioning Under a Comprehensive Decommissioning Program

The SDMP was developed by the staff, in response to the Commission's direction to develop a comprehensive strategy for NRC to deal with a number of contaminated sites, so that closure on cleanup issues could be attained in a timely manner.  In 1992, the staff developed the SDMP Action Plan to (a) identify criteria that would be used to guide the cleanup of sites, (b) state NRC's position on finality, (c) describe NRC's expectation that cleanup would be completed within 3–4 years, (d) identify guidance on site characterization, and (e) describe the process for timely cleanup on a site-specific basis.

Since development of the SDMP Action Plan, the staff has addressed the issues identified in the Action Plan, as follows.  The criteria for site cleanup and NRC's position on finality were codified in 10 CFR Part 20, Subpart E (LTR).  NRC's expectations regarding the completion of site decommissioning have been codified in 10 CFR 30.36, 40.42, 70.38, and  72.54.  Issues associated with site characterization have been addressed in NUREG-1575, Rev. 1 (MARSSIM, August 2000), and in NUREG-1757, Vol. 2, "Consolidated NMSS Decommissioning Guidance: Characterization, Survey, and Determination of Radiological Criteria, of the Consolidated NMSS Decommissioning Guidance" (September 2003).  The process for timely cleanup on a site-specific basis is addressed in NUREG-1757, "Consolidated NMSS Decommissioning Guidance."

Considering this, the staff has recently implemented the Commission's direction to eliminate the SDMP designation.  On June 17, 2004, the staff published a Notice in the *Federal Register* (69 *Federal Register* 33946) to announce that NRC has decided to eliminate the SDMP designation for sites and manage the SDMP sites as "complex sites" under a comprehensive decommissioning program.  Elimination of the SDMP designation and the discontinuance of the SDMP as a separate site listing is appropriate, because, as discussed above, the original intent of the SDMP and SDMP Action Plan (i.e., to achieve closure on cleanup issues so that cleanup could proceed in a timely manner) has been achieved.  The SDMP sites have been incorporated into a comprehensive decommissioning program that facilitates the cleanup of routine and complex sites in a manner that is consistent with the goals of the SDMP and SDMP Action Plan.

Viewed in the context of this comprehensive decommissioning program, which includes sites previously referred to as routine decommissioning sites, formerly licensed sites, SDMP sites, non-routine/complex sites, fuel cycle sites, and test/research and power reactors, the continued use of the SDMP list did not provide the same benefits that it did when it was first developed. The NRC staff believes the cleanup of these sites is now managed more effectively as part of this larger program.  As the SDMP sites are being managed as complex sites under this comprehensive program, the level of safety currently in place at SDMP sites has not been diminished.  In addition, as sites are identified and managed as complex sites, and as more sites are evaluated pursuant to the comprehensive decommissioning program, common problematic technical issues should be identified more easily, and resolutions to these issues should be implemented in a more consistent manner.

7.3     License Termination Rule Analysis Follow-up Actions

7.3.1   Analysis of License Termination Rule Issue on the Use of Intentional Mixing

In SECY-03-0069, "The Results of the LTR Analysis", staff identified the additional issue of the appropriateness of allowing intentional mixing of contaminated soil for meeting release criteria under the LTR.  The results of the staff's analysis of this issue were provided in SECY-04-0035 on March 1, 2004, and the Commission approved the recommendations of the staff in SRM-SECY-04-0035 on May 11, 2004.  The Commission approved allowing intentional mixing of soil to meet LTR release criteria in limited circumstances, on a case-by-case basis, and continuing the current practice of allowing intentional mixing for meeting waste acceptance criteria at offsite disposal facilities and for limited waste disposals.

7.3.2   Regulatory Issue Summary on License Termination Rule Analysis

On May 28, 2004, the staff issued RIS 2004-08, "Results of the License Termination Rule Analysis."  This RIS was the first LTR Analysis follow-up action of all the actions approved by the Commission in SRM-SECY-03-0069.  The purpose of the RIS was to inform licensees and stakeholders of NRC's analysis of the nine issues associated with implementing the LTR; the Commission's direction to date on how they can be addressed; schedule for future actions; and opportunities for stakeholder comment.  The RIS noted that stakeholder involvement would be an important part of developing the planned rulemaking and guidance.  In addition, early feedback was invited on the issues in the RIS.

7.3.3   Site-Specific Implementation of License Termination Rule Analysis Issues

During FY 2004 the staff began to implement, where appropriate, options approved by the Commission for the institutional control and realistic exposure scenario issues.  The progress toward implementation is summarized below, including site-specific examples.

Institutional Controls and Restricted Release

**Shieldalloy Metallurgical Corporation (SMC) site**:  Based on the Commission's approval of options and implementation actions in SRM-SECY-03-0069, and SMC's interest in using the Long-Term Control (LTC) license option, the staff developed interim guidance for the LTC license at the SMC site in New Jersey.  The interim guidance, provided to SMC in May 2004, include key concepts about the use of the LTC license as well as guidance on information to include in SMC's revised DP for institutional controls, engineering barriers, maintenance, and financial assurance.  As a follow-up to providing this guidance, a meeting was held in June 2004 to provide an opportunity to discuss the guidance with SMC and stakeholders.  SMC plans on revising its DP using the interim guidance and submitting the DP in FY 2005 for NRC review.  At the June meeting, State of New Jersey representatives discussed their June 25, 2004, letter to Chairman Diaz.  This letter raised several concerns with NRC's LTC license approach at the SMC site.  On September 9, 2004, Chairman Diaz provided New Jersey with a response to each concern and concluded that SMC should continue preparing its revised DP using the staff's interim guidance for the LTC license.

**AAR Manufacturing, Inc. (AAR)**: As discussed in SECY-03-0069, the staff has been working with AAR on the institutional control option of NRC monitoring and enforcing under a legal agreement and deed restriction. During FY 2004, the staff and AAR have been working on other technical issues related to the radiological survey and dose assessment that need to be resolved before further work can be done under the legal agreement option. The staff expects that this work will continue during FY 2005 as the staff develops its draft guidance for the institutional control issue.

**West Valley Demonstration Project (WVDP) site**: In March 2004, the staff met with DOE staff to discuss the scope and content of the DP for DOE's West Valley Demonstration Project (WVDP). During the discussion, NRC staff presented an update to the existing DP checklist for institutional controls. DOE will need to apply the risk-informed, graded approach to institutional controls as described in the LTR Analysis (SECY-03-0069) to determine which parts of the site need restrictions and the types of restrictions. The party that is ultimately responsible for institutional controls will be determined in the future, as a result of the ongoing process for developing the EIS.

Realistic Exposure Scenarios

During FY 2004, the staff started to implement the realistic exposure scenario approach approved by the Commission at the following nine decommissioning sites: Kiski Valley Water Pollution Control Authority (KVWPCA); FMRI, Inc. (Fansteel); SMC; AAR; Michigan Department of Natural Resources; SCA Services (SCA); DOW Chemical Co.; Cabot Performance Materials, Inc. (Cabot); and the WVDP. The KVWPCA and FMRI cases are discussed below and are of particular interest because they illustrate cases where the Commission approved policy has been tested at specific sites and either approved by the Commission or the Atomic Safety Licensing Board (ASLB).

**Kiski Valley Water Pollution Control Authority (KVWPCA) site**: In June 2004, the staff provided the results of its own dose assessment to support a recommendation to the Commission of no further decommissioning action (SECY-04-0102). The Commission approved the staff's recommendation for KVWPCA, including the application of the realistic scenario approach for this site. The dose assessment included a range of potential scenarios. An onsite, no action scenario was evaluated in which the contaminated lagoon is abandoned in place with no remedial actions performed. This was considered a reasonably foreseeable land use scenario. A removal scenario was also evaluated, in which the contaminated ash is excavated and removed to an offsite disposal facility. This also was considered reasonably foreseeable, based on the position of the Pennsylvania Department of Environmental Protection (PADEP) to excavate and remove the contaminated ash. This removal and offsite disposal scenario illustrates how the potential for offsite use should be evaluated consistent with the staff's approach discussed in the LTR Analysis RIS 2004-08. This offsite use approach was discussed in the RIS, in response to the Commission direction in SRM-SECY-03-0069. In addition to the reasonable foreseeable land use scenarios, the staff evaluated a pair of less likely cases, as assessment tools to bound the uncertainty associated with future land use.

**FMRI, Inc. (Fansteel) site**: The licensee proposed an industrial land use scenario for dose calculation purposes. The NRC staff reviewed the proposal and evaluated land use development in the area. The site is bounded on the north by the Port of Muskogee and industrial operations, on the east by the Arkansas River, on the south by US Highway 62, and on the west by the Muskogee Turnpike. In addition, there is a coal-fired power plant across the Arkansas River. The NRC staff confirmed the Port's interest in acquiring more land from FMRI for its operations. Based on this information, the staff concluded that industrial land use was appropriate for this site. However, the State of Oklahoma challenged that position, stating that the resident farmer should be used because there are other farms in the surrounding area. The State proposed that a recreational land use scenario be considered because the river and property across the river are used for recreation. After reviewing the issues and arguments, NRC's ASLB upheld the NRC staff's position. This decision is important because it is the first case that used the industrial scenario as a reasonably foreseeable land use, that had the approach challenged, and that was upheld by the ASLB.

7.3.4   Rulemaking and Supporting Guidance for License Termination Rule Analysis Issue on Preventing Future Legacy Sites

In SECY-03-0069, the staff recommended initiating a rulemaking and supporting guidance for measures to prevent future legacy sites. The Commission approved this action, including the development of a rulemaking plan. Upon further study, the staff informed the Commission on June 4, 2004, (COMSECY-04-0031) that it believed the objectives of a rulemaking plan (namely, to describe the regulatory problem and resolution, to propose a schedule, and to estimate resources) had already been included in SECY-03-0069. Consequently, the NRC staff's revised plan is to bypass the rulemaking plan that had been scheduled for completion in FY 2004 and proceed with developing the proposed rule. The Commission approved the staff's plans to proceed directly to the proposed rule stage (SRM-COMSECY-04-0031).

7.4   Decommissioning Program Evaluation

NRC's Strategic Plan for FY 2000-2005 identified a program evaluation, *Changes to the Decommissioning Process*, to be conducted in FY 2003. On September 29, 2003, the NRC staff completed its evaluation, included a summary in NRC's Annual Performance Report, and made the final report available on NRC's Decommissioning Website. In this report the staff evaluated the effectiveness of NRC's DWMEP Decommissioning Program and recommended future improvements. The staff evaluated overall program effectiveness with (a) NRC's Strategic Plan measures and targets, (b) NMSS Operating Plan accomplishments, and (c) the Office of Management and Budget (OMB) Program Assessment Rating Tool (PART). The staff used the PART questions as an independent methodology to systematically and comprehensively evaluate its program to identify areas of the program's effectiveness that might need further improvement. The staff also evaluated the effectiveness of 18 specific changes/improvements that were made to the program during the FY 2001–FY 2003 evaluation period. Independent reviews by the Commission and the Advisory Committee on Nuclear Waste were also used and add objectivity to the staff evaluations.

The staff concluded that the Decommissioning Program has been effective in meeting the Agency's strategic and performance measures and in closing/terminating sites after completion of decommissioning. The program also has effectively used many types of self assessments and program changes to improve the regulatory framework, decommissioning processes, internal program management processes, and openness. The staff believes these improvements have been useful and those that are ongoing should continue to be implemented. Although significant improvements have been completed, future improvements would be beneficial. In particular, the recommendations in the LTR Analysis (SECY-03-0069) to resolve the LTR policy issues, when implemented as directed by the Commission, offer potentially significant future improvements for the program. To complement these regulatory and policy improvements, the Program Evaluation makes additional recommendations that primarily would improve internal program management.

During FY 2004, an Improvement Plan was prepared that combines the implementation actions related to recommendations in the Program Evaluation with the Commission approved implementation actions related to the LTR Analysis in SECY-03-0069. Some of the Program Evaluation actions were started in FY 2004 [e.g., establishing a more comprehensive Decommissioning Program (SECY-04-0024)]; revising performance measures to be outcome oriented; and developing a risk-informed prioritization approach for managing site work).

## 8.    Resources

The total decommissioning program staff budget, for FY 2004 and FY 2005, is 74 full-time equivalents (FTEs) and 73 FTEs, respectively. These resource figures include: licensing casework directly related to decommissioning sites; inspections; project management and technical support for decommissioning power reactors, uranium mill tailings facilities and fuel cycle facilities; development of rules and guidance; environmental impact statements and environmental assessments; and research to develop more realistic analytical tools to support licensing and rulemaking activities such as controlling release of solid materials. These figures also include supervisory and non-supervisory indirect FTE, and training and travel associated with the decommissioning program.

## 9.    Reactor Decommissioning

### 9.1    Power Reactor Decommissioning

### 9.1.1   Power Reactor Decommissioning Process

The decommissioning process begins when a licensee decides to permanently cease operations. Several major steps make up the decommissioning process: notification; submittal and review of the Post-Shutdown Decommissioning Activities Report (PSDAR); submittal and review of the LTP; implementation of the LTP; and completion of decommissioning.

Notification

When the licensee has decided to permanently cease operations, it is required to submit a written notification to NRC. In addition, the licensee is required to notify NRC in writing once fuel has been permanently removed from the reactor vessel.

Post-Shutdown Decommissioning Activities Report

Prior to or within two years following cessation of operations, the licensee must submit a PSDAR. The PSDAR must include:

- A description and schedule for the planned decommissioning activities;
- An estimate of the expected costs; and

- A discussion that provides the means for concluding that the environmental impacts associated with the decommissioning activities will be bounded by appropriately issued EISs.

NRC will notice receipt of the PSDAR in the *Federal Register* and make the PSDAR available for public comment. In addition, NRC will hold a public meeting in the vicinity of the licensee's facility to discuss the PSDAR. NRC does not approve the PSDAR.

The licensee can not perform any major decommissioning activities until 90 days after NRC has received the PSDAR. After this period, the licensee can perform decommissioning activities as long as the activities do not:

- Foreclose release of the site for unrestricted use;
- Result in significant environmental impacts not previously reviewed; or
- Result in there no longer being reasonable assurance that adequate funds will be available for decommissioning.

In taking actions permitted under 10 CFR 50.59 following submittal of the PSDAR, the licensee must notify NRC in writing before performing any decommissioning activity inconsistent with, or making any significant schedule change from, those actions and schedules in the PSDAR.

License Termination Plan

Each power reactor must submit an application for termination of its license. The application must be accompanied or preceded by a LTP submitted for NRC approval. The LTP must include:

- A site characterization;
- Identification of remaining dismantlement activities;
- Plans for site remediation;
- Detailed plans for the final radiation survey;
- A description of the end use of the site, if restricted;
- An updated site-specific estimate of remaining decommissioning costs; and
- A supplement to the environmental report describing any new information or significant environmental change associated with the licensee's proposed termination activities.

In addition, the licensee must demonstrate that the applicable requirements of the LTR will be met.

NRC will notice receipt of the LTP and make the LTP available for public comment. In addition, NRC will hold a public meeting in the vicinity of the licensee's facility to discuss the LTP and the LTP review process. The review process is similar to that for material and fuel cycle licensees. The technical review is guided by NUREG-1700, "Standard Review Plan for Evaluating Nuclear Power Reactor License Termination Plans." The LTP is approved by license amendment.

Implementation of the License Termination Plan

Similar to material and fuel cycle facilities, NRC staff will inspect the licensee during decommissioning operations to ensure compliance with the LTP. These inspections will normally include in-process and confirmatory radiological surveys.

Decommissioning must be completed within 60 years of permanent cessation of operations unless otherwise approved by the Commission.

Completion of Decommissioning

At the conclusion of decommissioning activities the licensee will submit a final radiation survey report. NRC will terminate the license if it determines that:

- The remaining dismantlement has been performed in accordance with the approved LTP; and

- The final radiation survey and associated documentation demonstrates that the facility and site are suitable for release in accordance with the LTR.

9.1.2   Summary of FY 2004 Activities

Power reactor decommissioning activities include (a) project management for decommissioning power reactors and technical review responsibility for licensee submittals in support of decommissioning; (b) core inspections; and (c) supporting development of rulemaking and guidance.

NMSS currently has regulatory project management responsibility for 15 decommissioning power reactors. NRR has project management responsibility for two decommissioning reactors (Indian Point – Unit 1, Millstone – Unit 1) because extensive stakeholder interest in these sites makes it more efficient for NRR to retain, as a single point of contact, project management responsibilities for the permanently shutdown units. In addition, project management for three early demonstration reactors in decommissioning—Vallecitos, Nuclear Ship Savannah, and Saxton—remains with NRR. Plant status summaries for all decommissioning reactors are provided in Appendix A. During the past year, NMSS conducted reviews of the LTPs for Big Rock Point and Yankee Rowe. The staff expects to approve the Big Rock Point LTP in October 2004 and the Yankee Rowe LTP in April 2005. Table 9–1 provides a schedule for reactor decommissioning activities.

**Table 9–1**
**Power Reactors Undergoing Decommissioning**

| | Reactor | Location | PSDAR Submitted* | LTP Submitted | LTP Approved | Estimated License Term. | Site Summ. Pg. No. |
|---|---|---|---|---|---|---|---|
| 1 | Big Rock Point | Charlevoix, MI | 3/98 | 4/03 | 10/04 | 2012 | Page A-1 |
| 2 | Dresden – Unit 1 | Dresden, IL | 6/98 | TBD | TBD | TBD | Page A-2 |
| 3 | Fermi – Unit 1 | Newport, MI | 4/98 | 2005† | TBD | 2008 | Page A-3 |
| 4 | Haddam Neck – Connecticut Yankee | Meriden, CT | 8/97 | 7/00 | 11/02 | 2007 | Page A-4 |
| 5 | Humboldt Bay | Eureka, CA | 2/98 | 2007† | TBD | TBD | Page A-5 |
| 6 | Indian Point – Unit 1 | Buchanan, NY | 1/96 | TBD | TBD | TBD | Page A-6 |
| 7 | Lacrosse | La Crosse, WI | 5/91 | TBD | TBD | TBD | Page A-7 |
| 8 | Maine Yankee | Wiscasset, ME | 8/97 | 1/00 | 2/03 | 6/05 | Page A-8 |
| 9 | Millstone – Unit 1 | Waterford, CT | 6/99 | TBD | TBD | TBD | Page A-9 |
| 10 | Nuclear Ship Savannah | Newport News, VA | TBD | TBD | TBD | TBD | Page A-10 |
| 11 | Peach Bottom – Unit 1 | Delta, PA | 6/98 | 2012† | TBD | 2014 | Page A-11 |
| 12 | Rancho Seco | Sacramento, CA | 12/94 | 2005 | TBD | 2008 | Page A-12 |
| 13 | San Onofre – Unit 1 | San Clemente, CA | 12/98 | TBD | TBD | TBD | Page A-13 |
| 14 | Saxton | Saxton, PA | 1996 | 2/00 | 3/03 | 2004 | Page A-14 |
| 15 | Three Mile Island – Unit 2 | Harrisburg, PA | 2/79 | TBD | TBD | TBD | Page A-15 |

**Table 9–1**
**Power Reactors Undergoing Decommissioning (continued)**

| | Reactor | Location | PSDAR Submitted* | LTP Submitted | LTP Approved | Estimated License Term. | Site Summ. Pg. No. |
|---|---|---|---|---|---|---|---|
| 16 | Trojan | Rainier, OR | 1/96 | 8/99 | 2/01 | 6/05 | Page A-16 |
| 17 | Vallecitos | Pleasanton, CA | 7/66 | TBD | TBD | TBD | Page A-17 |
| 18 | Yankee Rowe | Greenfield, MA | 11/94 | 4/04 | 4/05 | 2021 | Page A-18 |
| 19 | Zion – Units 1 & 2 | Waukegan, IL | 2/00 | TBD | TBD | TBD | Page A-19 |

NOTES:

• Licensees submitted DPs (or equivalent) before 1996 and PSDARs after 1996.

* This is column includes the submission date (plus any revision dates) of the PSDAR or DP equivalent.

† This is an estimated date.

9.2     Research and Test Reactor Decommissioning

9.2.1   Research and Test Reactor Decommissioning Process

In general, the decommissioning process for research and test reactors and power reactors is the same (see Section 9.1.1).

9.2.2   Summary of FY 2004 Activities

NRR has project management and inspection responsibilities for 17 research and test reactors. Of these 17 research and test reactors, 13 have decommissioning orders or amendments; 3 are in "possession-only" status, either waiting for shutdown of another reactor at the site, or for removal of the fuel from the site by DOE; and 1 is preparing to submit a decommissioning amendment request. Further, 3 of the 13 research and test reactors with decommissioning orders or amendments, and 1 of the 3 research and test reactors in possession-only status, still have fuel in storage at the reactor. Table 9–2 identifies the decommissioning research and test reactors and provides the current status. Plant summaries for research and test reactors under NRR project management are provided in Appendix B.

**Table 9-2**
**Research and Test Reactors Undergoing Decommissioning**

| | Reactor | Location | Status | Estimated License Term. | Site Summ. Pg. No. |
|---|---|---|---|---|---|
| 1 | Cornell University – TRIGA | Ithica, NY | DECON Amendment | TBD | Page B-1 |
| 2 | Cornell University – ZPR | Ithica, NY | DECON Amendment | TBD | Page B-2 |
| 3 | Ford Nuclear Reactor | Ann Arbor, MI | DECON Amendment | TBD | Page B-3 |
| 4 | General Atomics – TRIGA Mark F | San Diego, CA | DECON Approved | TBD | Page B-4 |
| 5 | General Atomics – TRIGA Mark I | San Diego, CA | DECON Approved | TBD | Page B-5 |
| 6 | General Electric Co. – GETR | Sunol, CA | Possession Only | TBD | Page B-6 |
| 7 | General Electric Co. – VESR | Alameda, CA | Possession Only | TBD | Page B-7 |
| 8 | Manhattan College | Bronx, NY | DECON Approved | TBD | Page B-8 |
| 9 | NASA – Mockup | Sandusky, OH | DECON Approved | 2007 | Page B-9 |
| 10 | NASA – Plum Brook | Sandusky, OH | DECON Approved | 2007 | Page B-10 |
| 11 | University of Buffalo | Buffalo, NY | Possession Only | TBD | Page B-11 |
| 12 | University of Illinois | Urbana, IL | DECON Approved | TBD | Page B-12 |
| 13 | University of Virginia | Charlottesville, VA | DECON Approved | TBD | Page B-13 |
| 14 | University of Virginia – Cavalier | Charlottesville, VA | DECON Approved | TBD | Page B-14 |
| 15 | University of Washington | Seattle, WA | DECON Approved | TBD | Page B-15 |
| 16 | Veterans Administration | Omaha, NE | Operating License | TBD | Page B-16 |
| 17 | Westinghouse | New Stanton, PA | DECON Approved | 2005 | Page B-17 |

## 10.    Materials Facilities Decommissioning

### 10.1   Complex Site Decommissioning

As stated in Section 2, NRC has eliminated the SDMP designation for certain decommissioning facilities.  Instead, NRC will manage all materials decommissioning sites as "complex sites" under a comprehensive decommissioning program.  The SDMP designation will be used in this paper only to describe decommissioning activities which have taken place prior to June 17, 2004.  Currently, there are 43 complex decommissioning sites (see Table 10–1). Since last year's status report, five sites were removed from the complex site list:  (1) Babcock & Wilcox – Parks Township; (2) Envirotest Laboratories; (3) Molycorp, Inc. – York; (4) University of Wyoming; and (5) Watertown – GSA.

In Section 10.1.1, Table 10–1 identifies the clean-up criteria for each complex site as either LTR or SDMP Action Plan criteria.  The License Termination Rule (LTR) [Title 10 of the Code of Federal Regulations (CFR) Part 20, Subpart E] authorized two different sets of cleanup criteria—the concentration-based SDMP Action Plan criteria and the dose-based LTR criteria. Under the provisions of 10 CFR 20.1401(b), any licensee that submitted its DP before August 20, 1998, and received NRC approval of that DP before August 20, 1999, could use the SDMP Action Plan criteria for site remediation.  In the SRM on SECY-99-195, the Commission granted an extension of the DP approval deadline, for 12 sites, to August 20, 2000.  In September 2000, the staff notified the Commission that all 12 DPs were approved by the deadline.  All other sites must use the dose-based criteria of the LTR.

### 10.1.1 Complex Site Decommissioning Process

The decommissioning process is initiated by any one of the following conditions:

•       The license expires;

•       The licensee has decided to permanently cease principal activities at the entire site or in any separate building or outdoor area;

•       No principal activities have been conducted for a period of 24 months; or

•       No principal activities have been conducted for a period of 24 months in any separate building or outdoor area.

Several major steps make up the decommissioning process:  notification; submittal and review of the DP; implementation of the DP; and completion of decommissioning.

#### Notification

Within 60 days of the occurrence of any of the triggering conditions, the licensee is required to notify NRC of such occurrence and either begin decommissioning or, if required, submit a DP within 12 months of notification and begin decommissioning upon approval of the plan. Alternative schedules are authorized under the regulations, with NRC approval.

Decommissioning Plan

A DP must be submitted if required by license condition or if the procedures and activities necessary to decommission have not been previously approved by NRC and these procedures could increase potential health and safety impacts to workers or the public, such as in any of the following cases:

•       Procedures would involve techniques not applied routinely during clean up or maintenance operations;

•       Workers would be entering areas not normally occupied where surface contamination and radiation levels are significantly higher than routinely encountered during operation;

•       Procedures could result in significantly greater airborne concentrations than are present during operations; or

•       Procedures could result in significantly greater releases of radioactive material to the environment than those associated with operations.

The DP review process begins with an acceptance review. While primarily an administrative review, the acceptance review includes, but is not be limited to (a) completeness of the application; (b) legibility of drawings; (c) general adequacy of information; (d) justification for proprietary information; and (e) obvious technical inadequacies. The objective of the acceptance review is to verify that the application contains sufficient information before the staff begins an in-depth technical review. In addition, a limited technical review will be conducted. The purpose of the limited technical review is to identify significant technical deficiencies at an early stage, thereby precluding a detailed technical review of a technically incomplete submittal. At the conclusion of the acceptance review, the DP will either be accepted for detailed technical review or rejected and returned to the licensee with the deficiencies identified. For DPs proposing unrestricted release, a full technical review will be initiated after the successful conclusion of the acceptance review. The staff's review is guided by NUREG-1757, "Consolidated NMSS Decommissioning Guidance." The results of the staff's review will be documented in an Environmental Assessment (EA) and a Safety Evaluation Report (SER). The EA will be shared with the appropriate State, and State comments will be considered in finalizing the EA. The final EA must be summarized in the *Federal Register* in the form of a Finding of No Significant Impact (FONSI), provided an EIS is not necessary.

For reviews of DPs proposing restricted release, the review will be conducted in two phases. The first phase of the review will focus on the financial assurance (FA) and institutional control (IC) provisions of the DP. The review of the remainder of the DP will be initiated only after the staff is satisfied that the licensee's proposed IC & FA provisions will comply with the requirements of the LTR (10 CFR Part 20, Subpart E). The applicable portions of NUREG-1757, "Consolidated NMSS Decommissioning Guidance," will be used to guide this phase of the review. Phase II of the review will address all other sections of the technical review as guided by NUREG-1757 and will include the development of an EIS. Therefore, one of the first steps in Phase II is the publication of a Notice of Intent to develop an EIS. The basic EIS development steps are:

•       Notice of Intent;

•       Public scoping meeting;

- Preparation and publication of the scoping report;

- Preparation and publication of the draft EIS;

- Public comment period on the draft EIS including a public meeting;

- Preparation and publication of the final EIS; and

- Preparation and publication of the Record of Decision (ROD).

In parallel with the development of the EIS, the staff will develop a draft and final SER. The development of the draft SER will be coordinated with the development of the draft EIS so that any requests for additional information (RAIs) can be consolidated.

Regardless of whether an EA or EIS is developed, the staff structures its reviews so that the number of RAIs is minimized, without diminishing the technical quality or completeness of the licensee's ultimate submittal. For example, the staff will first develop a set of additional information needs and clarifications, including the bases for the additional information/ clarifications, and then meet with the licensee or responsible party to discuss the issues. This meeting will be noticed and conducted in accordance with NRC requirements for meetings open to the public. The results of the meeting will be documented in a meeting report. Any issues that can not be resolved during the meeting will be included in the formal RAI. In developing the final RAI, staff will document the insufficient or inadequate information submitted by the licensee and communicate what additional information is needed to address the identified deficiencies.

Following publication of the FONSI (for a DP involving an EA) or the ROD (for a DP involving an EIS), a license amendment will be issued approving the DP along with any additional license conditions found to be necessary as a result in the EA/EIS and/or the SER.

Implementation of the Decommissioning Plan

Following approval of the DP, the licensee must complete decommissioning in accordance with the approved DP within 24 months or apply for an alternate schedule. NRC staff will inspect the licensee during decommissioning operations to ensure compliance with the DP. These inspections will normally include in-process and confirmatory radiological surveys.

Completion of Decommissioning

As the final step in decommissioning, the licensee is required to:

- Certify the disposition of all licensed material, including accumulated wastes, by submitting a completed NRC Form 314 or equivalent information; and

- Conduct a radiation survey of the premises where licensed activities were carried out (in accordance with the procedures in the approved DP, if a DP is required) and submit a report of the results of the survey, unless the licensee demonstrates in some other manner that the premises are suitable for release in accordance with the LTR.

Licenses are terminated by written notice to the licensee when NRC determines that:

- Licensed material has been properly disposed of;

- Reasonable effort has been made to eliminate residual radioactive contamination, if present;

- Site meets the approved DP; and

- Radiation survey has been performed or other information submitted by the licensee that demonstrates that the premises are suitable for release in accordance with the LTR.

## 10.1.2 Summary of FY 2004 Activities

Material facilities decommissioning activities include (a) maintaining regulatory oversight of complex decommissioning sites, (b) undertaking financial assurance reviews, (c) examining issues and funding options to facilitate remediation of sites in non-Agreement States, (d) interacting with the EPA and ISCORS, (e) inspecting complex decommissioning sites, (f) maintaining the Computerized Risk Assessment and Data Analysis Lab (CRADAL), (g) conducting public outreach; (h) participating in International decommissioning activities, (i) conducting a program evaluation, and (j) participating in industry conferences and workshops.

- Activities associated with complex site decommissioning program include (a) review and approval of DPs, (b) conduct of pre-DP development meetings with licensees, (c) review of licensee FSSRs and conduct of confirmatory surveys, (d) conduct of in-process inspections, and (e) preparation of EAs and SERs. In FY 2004, the staff approved 4 DPs for the following sites: ABB Prospects, Inc.; Engelhard Minerals – Ohio; FMRI, Inc. (Fansteel); and NWI Breckenridge. The staff is currently reviewing DPs that were submitted in FY 2004 for the following sites: Dow Chemical Company; Michigan Department of Natural Resources; Pathfinder; SCA Services; and Westinghouse Electric Company (Hematite Facility).

- Staff routinely reviews financial assurance submittals for materials and fuel cycle facilities, and maintains a financial instrument security program. Approximately 50 financial assurance submittals were reviewed in FY 2004.

- The staff is currently preparing the annual update on issues and funding options to facilitate remediation of sites in non-Agreement States, which will be provided to the Commission in December 2004. For additional information on this subject, refer to Section 11.1.

- The staff continues to work with other Federal agencies, including EPA and DOE, through ISCORS, to address issues related to the radiation protection. ISCORS is nearing completion of its assessment of the origin, nature, and risk associated with radionuclides in sewage sludge from POTWs. The study has found that naturally occurring radionuclides are the primary contributor to radiation exposures. ISCORS is developing a Website that will include a catalog of parameters (such as inhalation and ingestion rates) used in dose modeling by different agencies and codes, to foster harmonization and consistency in the selection of parameters. ISCORS is also a forum for Federal agencies to discuss the wide range of radiation protection issues in decommissioning, including (a) standards for cleanup (EPA's "Federal Guidance for the General Public"), (b) use of institutional controls (c) cleanup criteria for radioactive dispersal device events, (d) disposition of solid materials, and (e) international initiatives related to protection of biota from ionizing radiation.

- CRADAL provides the staff with a high-performance computing capability that includes a platform to conduct intensive numerical calculations and parallel computing in support of licensing activities.

- One of the goals identified in NRC's Strategic Plan is to ensure openness in our regulatory process. The Strategic Plan identifies the development of communication plans for specific activities associated with the regulation of radiological decommissioning as a means to support the openness strategy. The staff continues to implement communication plans for all complex sites. Site-specific communication plans are useful tools to help ensure that the appropriate stakeholders are identified and contacted and focuses the staff on messages NRC wants to convey. One of the activities identified in the communication plans for each site is participation in public meetings to inform the public about major licensing actions. During the past year, the staff participated in public meetings for WVDP, Mallinckrodt Chemical Inc., Michigan Department of Natural Resources, SCA Services, and Pathfinder.

- The staff's participation in international activities is discussed in Section 5.

- The Decommissioning Program Evaluation is discussed in Section 7.

- The staff also participated in a number of industry conferences and workshops. Examples of conferences and workshops attended by the staff during the past year include Waste Management '04, American Nuclear Society conferences, and Health Physics Society meetings.

**Table 10–1**
**Current Complex Decommissioning Sites**

| | Name | Location | Date DP Submitted | Date DP Approved | Cleanup Criteria | Projected Removal | Site Summ. Pg. No. |
|---|---|---|---|---|---|---|---|
| 1 | AAR Manufacturing, Inc. | Livonia, MI | 10/97, Revised 7/05* | 5/98 2/06* | LTR-RES | 1/07 | Page C-1 |
| 2 | ABB Prospects, Inc. | Windsor, CT | 4/03 | 6/04 | LTR-UNRES | 12/07 | Page C-3 |
| 3 | Alliant Ordinance and Ground Systems, LLC (ATK) | Arden Hills, MN | 10/97 | 6/98 | Action-UNRES | 12/04 | Page C-4 |
| 4 | Augustana College | Sioux Falls, SD | NA | NA | LTR-UNRES | 9/04 | Page C-5 |
| 5 | Babcock & Wilcox (Shallow Land Disposal Area) | Vandergrift, PA | 6/01 | 5/05*+ | LTR-UNRES | 10/09 | Page C-6 |
| 6 | Battelle Columbus Laboratories | Columbus, OH | 8/00 | 2001 | LTR-UNRES | 12/05 | Page C-7 |
| 7 | Cabot Performance Materials, Inc. (Cabot) | Reading, PA | 11/02 | 3/05* | LTR-UNRES | 9/05 | Page C-8 |
| 8 | Curtis-Wright Cheswick | Cheswick, PA | TBD | TBD | LTR-UNRES | 12/08 | Page C-9 |
| 9 | Department of the Army | Fort McClellan, AL | 3/99 | 3/01 | LTR-UNRES | 6/05 | Page C-10 |
| 10 | Dow Chemical Company | Bay City, MI | 10/95, Revised 12/03 | 7/97 1/05* | LTR-UNRES | 4/06 | Page C-11 |

**Table 10–1**
**Current Complex Decommissioning Sites (continued)**

| Name | | Location | Date DP Submitted | Date DP Approved | Cleanup Criteria | Projected Removal | Site Summ. Pg. No. |
|---|---|---|---|---|---|---|---|
| 11 | Eglin Air Force Base | Walton County, FL | 8/03 | TBD | LTR-UNRES | 2005 | Page C-12 |
| 12 | Engelhard Minerals – Illinois | Great Lakes, IL | NA | NA | LTR-UNRES | 12/05 | Page C-13 |
| 13 | Engelhard Minerals – Ohio | Ravenna, OH | 06/03 | 03/04 | LTR-UNRES | 3/05 | Page C-14 |
| 14 | FMRI, Inc. (formerly Fansteel) | Muskogee, OK | 8/99, Revised 5/03 | 12/03 | LTR-UNRES | 2023+ | Page C-15 |
| 15 | Heritage Minerals | Lakehurst, NJ | 11/97 | 10/99 | Action-UNRES | 6/05 | Page C-17 |
| 16 | Homer Laughlin | Newell, WV | 1/95 | 1/95 | LTR-UNRES | TBD | Page C-18 |
| 17 | Jefferson Proving Ground (Department of Army) | Madison, IN | 8/99, Revised 6/02 | 10/02 | LTR-RES | TBD | Page C-19 |
| 18 | Kaiser Aluminum | Tulsa, OK | (Phase 1) 8/98, (Phase 2) 5/01 | 2/00 6/03 | Action-UNRES LTR-UNRES | 5/07 | Page C-20 |
| 19 | Kerr-McGee – Cimmarron | Cimarron, OK | 4/95 | 8/99 | Action-UNRES | 5/07 | Page C-21 |
| 20 | Kerr-McGee – Cushing Refinery Site | Cushing, OK | 8/98 | 8/99 | Action-UNRES | 12/05 | Page C-22 |
| 21 | Kerr McGee Tech. Center | Oklahoma City, OK | 4/01 | 6/03 | LTR-UNRES | 12/04 | Page C-23 |

**Table 10–1**
**Current Complex Decommissioning Sites (continued)**

| Name | | Location | Date DP Submitted | Date DP Approved | Cleanup Criteria | Projected Removal | Site Summ. Pg. No. |
|---|---|---|---|---|---|---|---|
| 22 | Kirtland Air Force Base | Albuquerque, NM | 11/02 | 1/03 | LTR-UNRES | 4/05 | Page C-24 |
| 23 | Kiski Valley Water Pollution Control Authority (KVWPCA) | Vandergrift, PA | NA | NA | LTR-UNRES | 11/04 | Page C-25 |
| 24 | Mallinckrodt Chemical, Inc. (Mallinckrodt) | St. Louis, MO | (Phase 1) 11/97 (Phase 2) 5/03 | 5/02 12/04* + | LTR-UNRES | 7/08 | Page C-26 |
| 25 | Michigan Department of Natural Resources | Kawkawlin, MI | 1/04 | 12/04* + | LTR-UNRES | 10/06 | Page C-27 |
| 26 | Molycorp, Inc. – Washington | Wash., PA | 6/99 | 8/00 | Action-UNRES | 10/06 | Page C-28 |
| 27 | NWI Breckenridge | Breckenridge, MI | 3/04 | 8/04 | LTR-UNRES | 12/04 | Page C-29 |
| 28 | Pathfinder | Sioux Falls, SD | 2/04 | 2/05* | LTR-UNRES | 4/06 | Page C-30 |
| 29 | Quehanna (formerly Permagrain Products, Inc.) | Media, PA | 4/98, Revised 3/03 | 7/98, 9/03 | Action-UNRES | 12/04 | Page C-31 |
| 30 | Royersford Wastewater Treatment Facility | Royersford, PA | TBD | TBD | LTR-UNRES | TBD | Page C-32 |

**Table 10–1**
**Current Complex Decommissioning Sites (continued)**

| | Name | Location | Date DP Submitted | Date DP Approved | Cleanup Criteria | Projected Removal | Site Summ. Pg. No. |
|---|---|---|---|---|---|---|---|
| 31 | Safety Light Corp. (SLC) | Bloomsburg, PA | 12/00 | 12/01 | LTR-UNRES | TBD | Page C-34 |
| 32 | Salmon River | Salmon, ID | TBD | TBD | LTR-UNRES | TBD | Page C-36 |
| 33 | SCA Services (SCA) | Kawkawlin, MI | 11/03 | 3/07*† | LTR-UNRES | 7/11 | Page C-37 |
| 34 | Shieldalloy Metallurgical Corp. (SMC) | Newfield, NJ | 2005* | 2006* | LTR-RES | 9/10 | Page C-38 |
| 35 | Stepan Chemical Company | Maywood, NJ | NA | NA | LTR-UNRES | 9/09 | Page C-39 |
| 36 | Superior Steel (formerly Superbolt) | Pittsburgh, PA | TBD | TBD | LTR-UNRES | TBD | Page C-40 |
| 37 | UNC Naval Products | New Haven, CT | 8/98 | 4/99 | LTR-UNRES | TBD | Page C-42 |
| 38 | Union Carbide Corp. | Lawrenceberg, TN (Buildings) (Soil) | 8/98 | 7/00 12/00 | Action-UNRES LTR-UNRES | 12/05 | Page C-43 |
| 39 | West Valley | West Valley, NY | 2005* | 2006* | LTR-UNRES** | TBD | Page C-44 |
| 40 | Westinghouse Electric Company | Blairsville, PA | NA | NA | LTR-UNRES | 12/04 | Page C-46 |
| 41 | Westinghouse Electric (Hematite Facility) | Jefferson City, MO | Phase 1 4/04 | Phase 1 12/05* | LTR-UNRES | 3/10 | Page C-47 |

34

**Table 10–1**
**Current Complex Decommissioning Sites (continued)**

| Name | | Location | Date DP Submitted | Date DP Approved | Cleanup Criteria | Projected Removal | Site Summ. Pg. No. |
|---|---|---|---|---|---|---|---|
| 42 | Westinghouse Electric, Waltz Mill | Madison, PA | 4/97 | 1/00 | LTR-UNRES | 8/05 | Page C-49 |
| 43 | Whittaker Corp. | Greenville, PA | 12/00, Revised 8/03 | TBD | LTR-UNRES | 9/05 | Page C-50 |

NOTES:

- The cleanup criteria identified in this table presents the staff's most recent information, but does not necessarily represent the current or likely outcome.

- Abbreviations used in this table include Action for SDMP Action Plan Criteria, LTR for LTR Criteria, RES for Restricted Use, and UNRES for Unrestricted Use.

\* This is an estimated date.

† The timeline for approving the DP is protracted due to (a) satisfying NEPA requirements, (b) conduct of public hearing, (c) multi-phase DP submittals, or (d) a combination of all the above.

\*\* The West Valley DP has not yet been submitted. The staff anticipates that West Valley DP will include plans to release a large portion of the site for unrestricted use, and the remainder of the site may have a perpetual license or be released with restrictions.

## 10.2    Uranium Recovery Facility Decommissioning

### 10.2.1 Uranium Recovery Facility Decommissioning Process

Decommissioning requirements for uranium recovery facilities are contained in 10 CFR 40.42 and supplemented by the criteria in Appendix A to Part 40.  Examples include the following:

- Criterion 5 provides ground-water protection requirements;
- Criterion 6 provides cover design requirements for uranium mill tailings impoundments and includes radiological criteria for decommissioning (Criterion 6(6));
- Criterion 6A requires a Commission-approved reclamation plan;
- Criteria 9 and 10 provide financial assurance requirements;
- Criterion 11 specifies site ownership requirements; and
- Criterion 12 specifies long-term surveillance requirements.

Guidance concerning the license termination process is contained in Appendix E of NUREG-1620, Rev.1, June 2003, "Standard Review Plan for the Review of a Reclamation Plan for Mill Tailings Sites Under Title II of the Uranium Mill Tailings Radiation Control Act of 1978." For the license termination of UMTRCA Title II sites under Agreement State jurisdiction, guidance is provided in Procedure SA-900 of the Office of State and Tribal Programs (STP).

### Role of Nuclear Regulatory Commission

In accordance with Section 83c of the Atomic Energy Act of 1954, as amended (AEA), NRC determines whether the licensee has met all applicable standards and requirements or whether a licensee-proposed alternative meets the standards.  This determination will involve NRC review of licensee submittals relative to the completion of decommissioning, reclamation, and, if necessary, groundwater cleanup.

In addition, the staff will review the site Long Term Surveillance Plan (LTSP) submitted by the custodial agency, for both NRC and Agreement State sites.  On NRC acceptance of the LTSP, NRC terminates the specific license and places the long-term care and surveillance of the site by the custodial agency under the general license provided at 10 CFR 40.28.

A final NRC responsibility is the determination of the final amount of long-term site surveillance funding.  Criterion 10 of Appendix A specifies a minimum charge of $250,000 (1978 dollars), revised to reflect inflation, which may be escalated on a site-specific basis because of surveillance and long-term monitoring controls beyond those specified in Criterion 10 of Appendix A.

### Role of Uranium Mill Licensees

Before license termination, licensees are required by license conditions to complete site decontamination, decommissioning, and surface and groundwater remedial actions consistent with decommissioning, reclamation, and groundwater corrective action plans.
Licensees must document the completion of these remedial actions in accordance with procedures developed by NRC.  This information will include a report documenting completion

36

of tailings disposal cell construction, as well as radiation surveys and other information required under 10 CFR 40.42.

Because the LTSP must reflect the remediated condition of the site, the licensee will work with the custodial agency in preparing the LTSP. Most likely, this coordination will involve supplying the custodial agency with appropriate documentation (e.g., as-built drawings) of the remedial actions taken and reaching agreements (formal or informal) with the custodial agency regarding the necessary surveillance control features of the site (e.g., boundary markers, fencing). It is the responsibility of the custodial agency to submit the LTSP to NRC for approval. However, the licensee may elect to help prepare the LTSP, to whatever degree is agreed upon between the licensee and the custodial agency.

Finally, the licensee provides the funding to cover long-term surveillance responsibilities in accordance with Criterion 10 of Appendix A. NRC will determine the final amount of this charge on the basis of final conditions at the site.

After termination of the existing license and transfer of the site and byproduct materials to the custodial agency, the remaining liability of the licensee extends solely to any fraudulent or negligent acts committed before the transfer to the custodial agency, as provided for in Section 83b(6) of the AEA.

Role of Custodial Agency

Section 83 of the AEA, as amended, states that before termination of the specific license, title to the site and byproduct materials should be transferred to (a) the DOE, (b) a Federal agency designated by the President, or (c) the State in which the site is located, at the option of the State. It is expected that the DOE will be the custodial agency for most, if not all, of the sites.

It is the responsibility of the custodial agency to submit the LTSP to NRC for review and acceptance. Provisions and activities identified in the final LTSP will form the bases of the custodial agency long-term surveillance at the site. The NRC general license in 10 CFR 40.28(a) becomes effective when the licensee's current specific license is terminated and the Commission accepts the LTSP. Custodial agencies are required, under 10 CFR 40.28(c)(1) and (c)(2), to implement the provisions of the LTSP.

The license termination process is discussed in more detail in Section E3.0 of NUREG-1620.

10.2.2  Summary of FY 2004 Activities

Uranium recovery decommissioning activities in the Division of Fuel Cycle Safety and Safeguards (FCSS) include (a) regulatory oversight of decommissioning uranium recovery (milling) sites; (b) review of site characterization plans and data; (c) review and approval of DPs; (d) preparation of EAs; (e) inspection of decommissioning, including confirmatory surveys; (f) decommissioning cost estimate reviews (including annual surety updates); and (g) oversight of license termination. The staff also reviews the DOE groundwater corrective-action plans and LTSPs for the Title I remediated mill sites and assists STP with review of Agreement-State uranium recovery site completion reports and inspections. At 13 of the Title I sites, NRC has concurred with DOE groundwater corrective action plans, and 7 other site plans are under review. Two sites currently are under active groundwater corrective action, and an additional

37

site will be active in the future.  The surface decommissioning at all Title I sites is complete.  In Section 10.2.3, Table 10–2 identifies the current Title II decommissioning sites and their status. Site summaries for the Title II decommissioning sites are provided in Appendix D.

During FY 2004, the staff completed over 60 licensing actions.  The most significant of these licensing actions include the following:

•       Approval of alternate concentration limits (ACLs) and runoff/erosion control measures for Rio Algom Mining Corporation (Rio Algom) – Lisbon;

•       Approval of the reclamation and DP for Plateau Resources, Inc.,– Shootaring Canyon;

•       Approval of a radon barrier for Pathfinder Mining Corporation – Lucky MC;

•       Resolution of a sulfate transport modeling issue for Petrotomics;

•       Approval of the mill demolition plan for Rio Algom – Ambrosia Lake;

•       Approval of a reclamation design for Umetco Minerals Corporation – Gas Hills; and

•       Approval of a rebaselined decommissioning cost estimate for COGEMA Mining, Inc. – Irigary/Christian.

In the SRM responding to SECY-03-0186, "Options and Recommendations for NRC Deferring Active Regulation of Ground-Water Protection at In Situ Leach Uranium Extraction Facilities," the Commission approved the staff's recommendations to defer such regulation (including decommissioning actions) to U.S. Environmental Protection Agency-authorized non-Agreement States and directed the staff to develop a RIS to obtain public comment about the staff's proposal before developing a MOU with each State.  On February 23, 2004, the staff issued RIS 2004-02 to request that, on a voluntary basis, addressees and other interested parties submit information pertaining to the proposed deferral.  On June 7, 2004, the staff issued RIS 2004-09 to inform addressees and other interested parties of (a) NRC's plans for the deferral and (b) the comments received in response to RIS 2004-02.  A notice of availability for each RIS was published in the *Federal Register*.  In addition, the staff has initiated groundwater protection program reviews for the States of Nebraska and Wyoming as discussed in SECY-03-0186.

On May 18–19, 2004, NRC staff participated in the National Mining Association/NRC Uranium Recovery Workshop in Denver, CO.  Over 100 individuals attended representing:  the DOE; EPA; State agencies; the industry; and members of the public.  The workshop was preceded on May 17, 2004, by public meetings with several licensees.

**Table 10–2**
**Current Decommissioning Title II Uranium Recovery Sites**

| | Name | Location | DP Approved | License Termination | Site Summ. Pg. No. |
|---|---|---|---|---|---|
| 1 | American Nuclear Corp. | Gas Hills, WY | 10/88, Revision 2005* | 2007 | Page D-1 |
| 2 | Bear Creek | Converse County, WY | 5/89 | 2004 | Page D-2 |
| 3 | COGEMA Mining, Inc. | Johnson & Campbell Counties, WY | 12/01 | 2007 | Page D-3 |
| 4 | ExxonMobil Highlands | Converse County, WY | 1990 | 2005 | Page D-4 |
| 5 | Homestake | Grants, NM | revised plan – 3/95 | 2015 | Page D-5 |
| 6 | Pathfinder – Lucky MC | Gas Hills, WY | revised plan – 6/96 | 2005 | Page D-6 |
| 7 | Pathfinder – Shirley Basin | Shirley Basin, WY | revised plan – 12/97 | 2007 | Page D-7 |
| 8 | Petrotomics | Shirley Basin, WY | 1989 | 2004 | Page D-8 |
| 9 | Rio Algom – Ambrosia Lake | McKinley Co., NM | 2003 (mill), 2004 (soil)* | 2008 | Page D-9 |
| 10 | Sequoyah Fuels Corp. | Gore, OK | 2006* | 2010 | Page D-10 |
| 11 | Sohio L-Bar | Seboyeta, NM | 5/89 | 2004 | Page D-11 |
| 12 | Umetco Minerals Corp. | East Gas Hills, WY | revised soil plan – 4/01 | 2006 | Page D-12 |
| 13 | United Nuclear Corp. | Church Rock, NM | 3/91, Revision 2005* | 2015 | Page D-13 |
| 14 | Western Nuclear, Inc. – Split Rock | Jeffrey City, WY | 1997 | 2007 | Page D-14 |

NOTE:

\*     This is the projected approval date.

## 10.3 Fuel Cycle Facility Decommissioning at Active Facilities

### 10.3.1 Fuel Cycle Facility Decommissioning Process

Some active facilities undergo partial decommissioning during operations. These facilities remain the responsibility of FCSS.

In general, the decommissioning process for fuel cycle facilities and complex material sites is the same (see Section 10.1.1). Project management responsibility for fuel cycle facilities resides in FCSS during licensee operations. Project management responsibility for decommissioning activities transfers to DWMEP for entire site decommissioning in support of license termination. However, the transfer from FCSS to DWMEP only occurs after the critical mass of material no longer remains at the site.

### 10.3.2 Summary of FY 2004 Activities

FCSS regulates facilities that mill and enrich uranium and fabricate it into fuel for use in nuclear reactors, and facilities that fabricate nuclear fuel that is a combination of uranium and plutonium oxides. Several types of fuel cycle facilities are licensed for the milling of uranium through its enrichment and fabrication into nuclear fuel used for nuclear power plants. These include uranium fuel fabrication facilities, uranium hexafluoride production (conversion) facility, gaseous diffusion enrichment facilities, and uranium milling facilities. Table 10–3 identifies the fuel cycle facilities with current decommissioning activities. Regulation of fuel cycle facilities is accomplished through a combination of regulatory requirements; licensing; safety oversight, including inspection, assessment of performance, and enforcement; operational experience evaluation; and regulatory support activities. Summaries of the decommissioning activities at fuel fabrication facilities are presented in Appendix E.

**Table 10–3**
**Fuel Cycle Facilities Undergoing Decommissioning**

|  | Name | Location | Status | Site Summ. Pg. No. |
|---|---|---|---|---|
| 1 | Framatome Richland | Richland, VA | Active | Page E-1 |
| 2 | General Atomics | San Diego, CA | Active | Page E-2 |
| 3 | Honeywell | Metropolis, IL | Active | Page E-3 |

## 11. FY 2005 Planned Programmatic Activities

### 11.1 Programmatic Initiatives

#### 11.1.1 Follow-up Actions to Implement Decommissioning Program Evaluation Recommendations

Follow-up actions to the Decommissioning Program Evaluation are also planned for FY 2005. Examples include making decommissioning Website enhancements; holding training on NUREG-1757 for dose modeling and risk-informed, performance-based approach application; developing a resource tracking system; considering options for sharing decommissioning approaches and lessons; considering options for and feasibility of independent review of NRC's decommissioning program; and considering use of incentives to facilitate licensee decommissioning.

#### 11.1.2 Follow-up Actions for Sites with Inadequate Financial Assurance

SECY-03-0198, "Progress and Future Plans for Sites Identified in SECY-02-0079 with Inadequate Financial Assurance," summarized staff's progress and plans for several sites identified with inadequate financial assurance and recommended continuing the financial program. SRM-03-0198 approved staff's plan to continue the financial program and update the Commission annually. Although both the staff and Commission considered including the update as part of the annual update of the Comprehensive Decommissioning Program, the staff believes that continuing a separate annual update is preferable considering the financial information included in the paper is sensitive and therefore needs to be decoupled from the overall annual report that is publically available. As a result, the staff plans on providing the Commission with an annual update in December 2004. This report will include (a) a site table that summarizes progress for each of the existing 13 sites by identifying issues resolved, issues pending, and the path forward; (b) another table that identifies the effects of the innovative approaches to implementing the LTR on the Group II sites, as directed by the Commission; (c) any new sites added to the financial program; and (d) other issues such as consideration of licensing non-licensed sites.

#### 11.1.3 Prepare Draft Update of NUREG-1757 for Comment

The NRC staff finalized its "Consolidated NMSS Decommissioning Guidance" (NUREG-1757, Vols. 1–3) in 2003, and the NRC staff intended periodic updates. In FY 2005, the staff plans to evaluate changes to the guidance that may be needed, and initiate modifications or supplements to the guidance, as appropriate. The staff plans to make available for public review and comment any proposed revisions to the guidance.

#### 11.1.4 Uranium Recovery Actions Requiring Consultation with the Commission in FY 2005

The staff plans to consult with the Commission in June 2005 concerning the final MOUs with the States of Nebraska and Wyoming regarding the deferral of active groundwater regulation at ISL facilities in those States.

11.2    Rulemaking

The staff plans on conducting numerous activities in FY 2005 to implement the Commission approved LTR Analysis follow-up actions.  For the rulemaking and supporting guidance on measures to prevent future legacy sites (changes to financial assurance and licensee operations), the staff plans on developing a technical basis and beginning to prepare the proposed rule and draft guidance, which would be issued for public comment during FY 2006. The staff also will develop draft guidance for public comment in FY 2005 for the following LTR Analysis issues:  (a) institutional controls/restricted release; (b) realistic exposure scenarios; (c) onsite disposal; (d) control of the disposition of solid materials; and (e) intentional mixing of soil.  Finally, the staff plans on developing revised inspection procedures and enforcement guidance by using a risk-informed approach to enhance monitoring, reporting, and remediation at operating facilities to reduce the likelihood of future decommissioning problems or sites with insufficient funds for cleanup and decommissioning.

11.3    International Activities

Many of the DWMEP international activities discussed previously are ongoing arrangements with the international community.  These include the following.

11.3.1 Support to the International Atomic Energy Agency

NRC staff will continue to provide support to the IAEA in the following areas:

- The Waste Safety Standards Committee (WASSC) reviews the IAEA regulatory criteria development program for waste safety, including decommissioning.  The director of DWMEP also sits on the WASSC, as the U.S. representative.  In FY 2005 and into the future, the WASSC will continue to promote waste safety and revisit past decommissioning guidance, to determine whether the criteria and guidance need to be revised to address improvements in technology or modifications in the understanding of the impacts of sites and facilities that have been released from regulatory control.  For example, the IAEA has recently separated decommissioning from predisposal waste management as a program area.  Thus, activities in FY 2005 will include the development of separate safety requirements and guidance in this particular area.  The WASSC meets biannually in the spring and the fall.

- The Commission on Safety Standards is an oversight body, which provides a unified review and approval of regulatory documents forwarded by the various IAEA safety standards committees (including WASSC).  It also meets biannually and would provide a final technical and programmatic approval of the regulatory documents.

- Decommissioning documents expected for review by the two committees include the following:
  — a DS-333, "Decommissioning of Nuclear Facilities" (scheduled for publication in June 2006); and
  — DS-332, "The Removal of Sites and Buildings from Regulatory Control upon the Termination of Practices" (scheduled for publication in May 2005).

- The Joint Convention on the Safety of Spent Fuel Management and on the Safety of Radioactive Waste Management included all of waste management, including decommissioning. During FY 2005, the interagency writing group will continue preparation of the U.S. National Report, and the NRC staff will complete the internal NRC review and approval process. The National Report is due to the IAEA by October 15, 2005.

- NRC staff will prepare briefing book materials for senior NRC management for their participation in the annual IAEA General Conference and Board of Governors' meeting, which is usually held in the fall of each year.

- DWMEP staff will be asked to review various documents, such as Action Plans, that are the product of the previous year's IAEA General Conference. These may include radioactive waste management and in particular decommissioning.

## 11.3.2 Support to the Organization for Economic Cooperation and Development's Nuclear Energy Agency

As in the past years, decommissioning issues are specifically addressed by a standing subcommittee of the OECD/NEA, the WPDD. In the past, meetings have been in conjunction with other ongoing NEA activities such as the meetings of the Radioactive Waste Management Committee or with NEA decommissioning conferences.

## 11.3.3 Bilateral and Trilateral Exchanges with Other Countries

Currently, there are two standing exchanges with other countries: the bilateral exchange with the French DGSNR and a trilateral exchange with Mexico and Canada. Again, decommissioning is one of the many topics raised and discussed. The bilateral exchange with the French takes place twice a year, once in the United States and once in France. The trilateral exchange takes place annually.

Hosting Foreign Assignees and Providing Reciprocal Assignments

DWMEP expects to host a foreign assignee interested in decommissioning from Taiwan in July 2006 or January 2007. An assignee from the People's Republic of China is scheduled to terminate his assignment in October 2004. An assignee from Spain is under consideration for assignment to the decommissioning area in the near future.

Other Activities

DWMEP will continue to support assistance requests from the OIP, as needed. The types of support activities performed in the past are described in Section 4.

DWMEP plans to participate in the Decommissioning Workshop in Moscow, Russia, in September 2005. This is part of an NRC initiative to provide assistance to other countries in a bilateral spirit of cooperation. A similar workshop was held in Taipei, Taiwan in March 2004.

The staff will also provide support to international conferences, such as the International Conference on Environmental Management, which is held on a 2-year (biennial) basis at various sites with interest in environmental restoration and decommissioning.

# Appendix A

# Site Summaries for
# Power Reactors Undergoing
# Decommissioning

# BIG ROCK POINT

## 1.0 SITE IDENTIFICATION

Location:           Charlevoix, MI
License No.:        DPR-6
Docket No.:         50-155
License Status:     Permanently shutdown
Project Manager:    Jim Shepherd

## 2.0 SITE STATUS SUMMARY

The plant was permanently shut down on August 29, 1997. Fuel was transferred to the spent fuel pool by September 20, 1997. On September 19, 1997, the Consumers Energy Company (CE) submitted a PSDAR that identified decommissioning activities commencing in September 1997, and concluding in September 2002. The licensee selected the DECON option. On March 26, 1998, CE submitted a revised PSDAR that showed conclusion of decommissioning about August 2005. Dry fuel storage will continue through about 2012, depending on when the DOE accepts fuel. CE is currently decommissioning the site in accordance with the PSDAR.

All fuel was transferred to the independent spent fuel storage installation (ISFSI) and the spent fuel pool has been drained and cleaned. The reactor vessel was shipped to Barnwell on October 7, 2003.

On April 1, 2003, CE submitted its LTP. By this plan, CE will release those parts of the site not needed for ISFSI operation at the completion of the remediation project. After fuel is removed from the site, the ISFSI will be decommissioned and the license terminated. The reactor head was shipped to Envirocare on May 28, 2003. NRC sent a RAI to CE on February 13, 2004. The licensee responded on July 1, 2004, with specific answers and a revision to the LTP. The staff expects to have the LTP approved in October 2004.

## 3.0 MAJOR TECHNICAL OR REGULATORY ISSUES

Contaminants at the site include uranium and decay products, and fission products. Groundwater contamination is non-uniformly distributed at the site because of a dry, silty clay layer that underlies only the south part of the site. Boundaries between the geologic units are only approximated because of limited subsurface data; additional data may be necessary to determine the extent of contamination. Reported concentrations in ground water are low, generally less than the minimum detectable activity (MDA) except for tritium. Soil contamination is also generally below MDA.

There is some public interest about the decommissioning of this site. The primary parties are the State of Michigan and the City Councils of surrounding areas. CE has an effective public outreach program and open communication with these parties.

## 4.0 ESTIMATED DATE FOR LICENSE TERMINATION    2012

# DRESDEN – UNIT 1

## 1.0  SITE IDENTIFICATION

| | |
|---|---|
| Location: | Dresden, IL |
| License No.: | DPR-2 |
| Docket No.: | 50-0010 |
| License Status: | Permanently shutdown |
| Project Manager: | John Hickman |

## 2.0  SITE STATUS SUMMARY

The plant shut down in October 1978 and is currently in SAFSTOR.  The DP was approved in September 1993.  No significant dismantlement activities are underway.  Asbestos removal, isolation of Unit 1 from Units 2 and 3, and general radiation cleanup activities are complete or in progress.  The licensee will dismantle Unit 1 at the same time as the other two units onsite, which is expected no earlier than 2011.  The licensee submitted an updated PSDAR on June 1, 1998.  The PSDAR public meeting was held on July 23, 1998.

## 3.0  MAJOR TECHNICAL OR REGULATORY ISSUES

The licensee is using the Holtec HISTAR 100 dual purpose cask and the HISTORM concrete overpack to store spent fuel.  Casks have been loaded with Unit 1 spent fuel from the Unit 2 spent fuel pool, along with Unit 2 spent fuel, to address the Unit 2 spent fuel storage issue.  In January 2002, the licensee completed transferring fuel from the Unit 1 spent fuel pool to dry storage.

## 4.0  ESTIMATED DATE FOR LICENSE TERMINATION     TBD

# FERMI – UNIT 1

## 1.0 SITE IDENTIFICATION

Location:            Newport, MI
License No.:         DPR-9
Docket No.:          50-16
License Status:      Permanently shutdown
Project Manager:     Ted Smith

## 2.0 SITE STATUS SUMMARY

The licensee's initial stage of decommissioning is complete, and bulk sodium has been removed from the site.  There is no spent fuel onsite and the facility is currently in SAFSTOR condition.  The licensee is currently performing occupational safety enhancement activities; concentrating in non-radioactive areas, such as asbestos removal, and trace sodium cleanup.  The trace sodium remediation effort is about 50 percent complete.  The facility will be dismantled under the provisions of 10 CFR 50.59.  The licensee plans to submit an LTP in 2005.

## 3.0 MAJOR TECHNICAL OR REGULATORY ISSUES

None

## 4.0 ESTIMATED DATE FOR LICENSE TERMINATION    2008

# HADDAM NECK – CONNECTICUT YANKEE

## 1.0  SITE IDENTIFICATION

Location:            Meriden, CT
License No.:         DPR-61
Docket No.:          50-213
License Status:      Permanently shutdown
Project manager:     Ted Smith

## 2.0  SITE STATUS SUMMARY

Steam generators, reactor coolant pumps, the pressurizer, the reactor vessel, and shield wall blocks from the Reactor Building shielding have been disposed offsite.  Demolition of the administration and turbine buildings began in spring 2004.  There are 1016 spent fuel assemblies and 18 canisters of greater than Class C waste stored in the spent fuel pool.  The ISFSI is currently operational, and the SNF loading campaign is underway and is scheduled to be completed in early 2005.  The licensee is modifying the spent fuel building in order to use of a second transfer cask which will expedite movement of spent fuel to the ISFSI.

The licensee is reevaluating their preliminary groundwater findings using a revised packer test procedure.  Removal of source term material near the tank farm began in June 2004 and is scheduled to be complete by November 2004.

The staff completed its review of the LTP and issued its safety evaluation on November 25, 2002.  The LTP was challenged by the Citizens Awareness Network, and determined in CY's favor on November 24, 2003.  In a change to NRC-approved decommissioning approach, the licensee will now be removing all structures from the site down to four feet below grade.

## 3.0  MAJOR TECHNICAL OR REGULATORY ISSUES

None

## 4.0  ESTIMATED DATE FOR LICENSE TERMINATION     2007

# HUMBOLDT BAY

## 1.0 SITE IDENTIFICATION

Location:          Eureka, CA
License No.:       DPR-7
Docket No.:        50-133
License Status:    Permanently shutdown
Project Manager:   Bill Huffman

## 2.0 SITE STATUS SUMMARY

The plant was shut down in July 1976 and has been in SAFSTOR ever since. A DP was approved in July 1988. Subsequent to the 1996 decommissioning rule, the licensee converted the DP into its Defueled Safety Analysis Report which is now updated every two years. A PSDAR was issued by the licensee in February 1998. The plant is currently in SAFSTOR with incremental decommissioning activities ongoing. Decommissioning work at Humboldt Bay involves recently completed asbestos removal, currently in progress systems and structures radiological characterization, and future work on reactor and internals activation analysis, low-level waste (LLW) management plan development, developing of a work, cost, and scheduling process, and the developing of a facilities and staffing plan. This work phase will likely continue until a decision is made on accelerated decommissioning. The licensee's estimated remaining decommissioning costs are approximately $270 million. The licensee currently has approximately $173 million in its decommissioning funds with an additional $41 million to be collected over the next 2 years.

## 3.0 MAJOR TECHNICAL OR REGULATORY ISSUES

The licensee submitted an ISFSI application in December 2003. The ISFSI dry storage cask will be unique due to the short length of the Humboldt fuel assemblies. Furthermore, the casks will be stored below-grade to accommodate regional seismicity issues, security concerns, and site boundary dose limits. Review and approval of the ISFSI application by NRC is ongoing. If the ISFSI application is approved, a decision will then be made on whether to proceed with ISFSI construction.

Recent inspection and inventory of the spent fuel pool has revealed numerous fuel pin fragments presumably from failed fuel assemblies processed while the plant was operational. Furthermore, the licensee is trying to account for several fuel pin segments that should be stored in the spent fuel pool but cannot be located.

## 4.0 ESTIMATED DATE FOR LICENSE TERMINATION    TBD

# INDIAN POINT – UNIT 1

## 1.0 SITE IDENTIFICATION

Location:           Buchanan, NY
License No.:        Provisional License DPR-5
Docket No.:         50-3
License Status:     Permanently shutdown
Project Manager:    Michael Webb

## 2.0 SITE STATUS SUMMARY

The plant was shutdown in October 1974.  Some decommissioning work associated with spent fuel storage was performed from 1974 through 1978.  The order approving SAFSTOR was issued in January 1996.  The PSDAR public meeting was held on January 20, 1999.  The licensee plans to decommission Unit 1 with Unit 2, which is currently in operation.  The licensee does not plan to begin active decontamination and decommissioning until 2013, when the IP2 license expires.

Since purchasing the Indian Point facility, Entergy has been reviewing its long-term spent fuel storage options for Unit 1, but has not finalized its plans.

No date has been set for the transfer of project management from the Office of Nuclear Reactor Regulation (NRR) to the Office of Nuclear Material Safety and Safeguards (NMSS).

## 3.0 MAJOR TECHNICAL OR REGULATORY ISSUES

There are currently no major regulatory issues associated with Indian Point 1.  The staff is reviewing one licensing action, a revision to the Quality Assurance Program, that affects all Entergy facilities and is not unique to Indian Point 1.

## 4.0 ESTIMATED DATE FOR LICENSE TERMINATION     TBD

# LACROSSE

## 1.0 SITE IDENTIFICATION

Location:            La Crosse, WI
License No:          DPR-45
Docket No.:          50-409
License Status:      Permanently shutdown
Project Manager:     Bill Huffman

## 2.0 SITE STATUS SUMMARY

The plant was shut down on April 30, 1987. The SAFSTOR DP was approved August 7, 1991. The DP is considered the PSDAR. The PSDAR public meeting was held on May 13,1998. Limited and gradual dismantlement is currently underway. The owner is a member of the Private Fuel Storage LLC seeking a license to build and operate an ISFSI on the reservation of the Skull Valley Band of Goshute Indians west of Salt Lake City, Utah. The owner has no immediate plans for an onsite ISFSI. Active decommissioning is not expected to commence until all spent fuel is removed from the spent fuel pool and placed in dry storage or transferred to a Federal repository.

## 3.0 MAJOR TECHNICAL OR REGULATORY ISSUES

None

## 4.0 ESTIMATED DATE FOR LICENSE TERMINATION    TBD

# MAINE YANKEE

## 1.0 SITE IDENTIFICATION

Location:              Wiscasset, ME
License No.:           DPR-36
Docket No.:            50-309
License Status:        Permanently shutdown
Project Manager:       John Buckley

## 2.0 SITE STATUS SUMMARY

The plant was shutdown on December 6, 1996. Certification of permanent cessation of operations was submitted on August 7, 1997. The PSDAR was submitted on August 27, 1997. The LTP was submitted on January 13, 2000 and subsequently revised on June 1 and August 13, 2001. The LTP was approved on February 28, 2003.

Spent fuel transfer (1432 fuel assemblies in 60 casks) from the spent fuel pool to the onsite ISFSI began in August 2002 and was completed in February 2004. All decommissioning activities at the site are scheduled to be completed in March 2005.

## 3.0 MAJOR TECHNICAL OR REGULATORY ISSUES

The licensee has submitted a license amendment request to shrink the Part 50 license to the ISFSI island upon completion of the decommissioning. The request is under staff review.

## 4.0 ESTIMATED DATE FOR LICENSE TERMINATION

Decommissioning complete 6/05, retain Part 50 license until fuel is removed from ISFSI

# MILLSTONE – UNIT 1

## 1.0 SITE IDENTIFICATION

Location:          Waterford, CT
License No:        DPR-21
Docket No.:        50-245
License Status:    Permanently shutdown
Project Manager:   Alan Wang

## 2.0 SITE STATUS SUMMARY

Unit 1 was shut down on November 4, 1995, and transfer of the spent fuel to the pool was completed on November 19, 1995. On July 17, 1998, the licensee decided to cease operations. Certifications per 10 CFR Part 50.82(a) were submitted July 21, 1998. The owner's current plan is to leave the plant in SAFSTOR until the Unit 2 license expires. The owner submitted its required PSDAR on June 14, 1999, and has chosen a combination of the DECON and SAFSTOR options. NRC conducted public meetings in Waterford, CT, on the decommissioning process on February 9, 1999, and on the PSDAR on August 25, 1999. Owner responsibility for the Millstone site was transferred from Northeast Utilities to Dominion Nuclear Connecticut on March 31, 2001. Unit 1 has established a spent fuel pool island including those systems required to support safe storage of spent fuel. The balance of systems not required to support the facility have been abandoned. Irradiated reactor vessel components not able to eventually being disposed of with the reactor vessel have been removed. The reactor cavity and vessel will be drained and abandoned with a radiation shield installed to limit dose to workers.

No date has been set for the transfer of project management from NRR to NMSS.

## 3.0 MAJOR TECHNICAL OR REGULATORY ISSUES

None

## 4.0 ESTIMATED DATE FOR LICENSE TERMINATION     TBD

# NUCLEAR SHIP SAVANNAH

## 1.0 SITE IDENTIFICATION

Location:            Newport News, VA
License No.:         NS-1
Docket No.:          50-238
License Status:      Permanently shutdown
Project Manager:     Al Adams

## 2.0 SITE STATUS SUMMARY

The reactor is currently in SAFSTOR. All fuel has been removed from the ship. The Nuclear Ship (NS) Savannah is moored in the Maritime Administration Reserve Fleet in the James River, Virginia. As needed, the NS Savannah is towed into dry dock for hull maintenance. Because the reactor is portable, the location of decommissioning has not been determined. There are no plans to transfer NRR project management to NMSS project management.

## 3.0 MAJOR TECHNICAL OR REGULATORY ISSUES

The licensee is exploring the possibility of obtaining funding for total decommissioning and disposal of the NS Savannah.

## 4.0 ESTIMATED DATE FOR LICENSE TERMINATION    TBD

# PEACH BOTTOM – UNIT 1

## 1.0 SITE IDENTIFICATION

Location:            Delta, PA
License No.:         DPR-12
Docket No.:          50-171
License Status:      Permanently shutdown
Project Manager:     Kristina Banovac

## 2.0 SITE STATUS SUMMARY

The facility has been permanently shutdown since October 31, 1974 and is currently in a SAFSTOR condition.  The licensee will maintain its facility in SAFSTOR until 2010 and submits its LTP in 2012.  Spent fuel has been removed from the site.  The PSDAR meeting was held on June 29, 1998.  Final decommissioning is not expected until 2015 when Units 2 and 3 are scheduled to shut down.

## 3.0 MAJOR TECHNICAL OR REGULATORY ISSUES

None

## 4.0 ESTIMATED DATE FOR LICENSE TERMINATION    2014

# RANCHO SECO

## 1.0 SITE IDENTIFICATION

Location:          Sacramento, CA
License No.:       DPR-54
Docket No.:        50-312
License Status:    Permanently shutdown
Project Manager:   John Hickman

## 2.0 SITE STATUS SUMMARY

The plant was shut down in June 1989. The SAFSTOR DP was approved in March 1995. The licensee revised its DP to use an incremental dismantlement approach. Currently, the licensee is dismantling the secondary side of the plant. Wastes generated during decommissioning will be shipped to Envirocare. In July 1999, the owner decided to continue in DECON, with the goal of completing the decommissioning by 2008. On October 4, 1991, the owner submitted a site-specific Part 72 ISFSI application using the VECTRA NUHOMS-MP187 dual purpose cask design. The license was granted on June 30, 2000. The owner has transferred all of the spent fuel from the pool to the onsite ISFSI. The LTP is scheduled to be submitted in 2005.

## 3.0 MAJOR TECHNICAL OR REGULATORY ISSUES

None

## 4.0 ESTIMATED DATE FOR LICENSE TERMINATION    2008

# SAN ONOFRE – UNIT 1

## 1.0  SITE IDENTIFICATION

| | |
|---|---|
| Location: | San Clemente, CA |
| License No.: | DPR-13 |
| Docket No.: | 50-206 |
| License Status: | Permanently shutdown |
| Project Manager: | Bill Huffman |

## 2.0  SITE STATUS SUMMARY

The plant was shut down in November 1992.  The licensee submitted an updated PSDAR on December 15, 1998.  The facility transitioned from SAFSTOR in 1999 and is now in DECON. Significant dismantlement is currently underway.  The licensee has completed demolition of the Emergency Diesel Generator building, the Control Building, and Administration Building. Dismantlement and removal of the electrical generator and main turbine is also complete.  The licensee has completed reactor pressure vessel internal segmentation and cutup.  The reactor internals abrasive cutting media has been sent offsite for disposal.  Most of the Containment Sphere Enclosure Building has been dismantled and most of the large reactor system components have been removed including the reactor pressure vessel, pressurizer and steam generators.  The control room has been relocated and Unit 1 has established its spent fuel pool island concept with the rest of the Unit 1 facility cold and dark.  Major security modifications to isolate Units 2 and 3 from Unit 1 are complete.  The steam generators and pressurizer have been shipped to disposal.  The licensee was unable to make arrangements for shipping the reactor pressure vessel to disposal due to the size and weight of the vessel and shipping package.  The licensee plans to store the vessel onsite for the foreseeable  as long as licensed activities are ongoing.  The licensee is currently transferring Unit 1 spent fuel to an onsite generally licensed ISFSI and expects to complete the Unit 1 spent fuel transfer by September 2004.

## 3.0  MAJOR TECHNICAL OR REGULATORY ISSUES

None

## 4.0  ESTIMATED DATE FOR LICENSE TERMINATION     TBD

# SAXTON

## 1.0 SITE IDENTIFICATION

| | |
|---|---|
| Location: | Saxton, PA |
| License No.: | DPR-4 |
| Docket No.: | 50-146 |
| License Status: | Permanently shutdown |
| Project Manager: | Al Adams |

## 2.0 SITE STATUS SUMMARY

The plant was shut down in May 1972, and in February 1975, was placed in SAFSTOR until 1986 when phased dismantlement began with removal of support buildings, contaminated soil, and some material in the containment.  The owner submitted a DP in 1996, which became the PSDAR.  All spent fuel has been removed from the site.  NRC approved an amendment request in 1998 to allow dismantlement under 10 CFR 50.59.  The reactor vessel with internals, steam generator, and pressurizer have been shipped to Barnwell for disposal.  The owner submitted a LTP in February 1999, but had to resubmit the plan in February 2000 to provide sufficient information for an acceptance review.  NRC approved the LTP on March 28, 2003.  The site is in DECON, and the owner expects to complete decommissioning so the license can be terminated in the fourth quarter of 2004 and the site restored by the first quarter of 2005.

No date has been set for the transfer of project management from NRR to NMSS.

## 3.0 MAJOR TECHNICAL OR REGULATORY ISSUES

None

## 4.0 ESTIMATED DATE FOR LICENSE TERMINATION    2004

# THREE MILE ISLAND – UNIT 2

## 1.0  SITE IDENTIFICATION

Location:             Harrisburg, PA
License No.:          DPR-73
Docket No.:           50-320
License Status:       Permanently shutdown
Project Manager:      Bill Huffman

## 2.0  SITE STATUS SUMMARY

The operational accident occurred in March 1979.  The plant defueling was completed in April 1990.  Post Defueling Monitored Storage was approved in 1993.  There is no significant dismantlement underway.  The plant shares equipment with the operating TMI–Unit 1.  TMI-1 was sold to Amergen in 1999.  GPU Nuclear retains the license for TMI-2 and contracts to Amergen for maintenance and surveillance activities.  All spent fuel has been removed except for some debris in the nuclear steam supply system.  The removed fuel is currently in storage at Idaho National Engineering Laboratory.  DOE has taken title and possession of the fuel debris.  The licensee plans to actively decommission Unit 2 in parallel with the decommissioning of TMI–Unit 1.

## 3.0  MAJOR TECHNICAL OR REGULATORY ISSUES

None

## 4.0  ESTIMATED DATE FOR LICENSE TERMINATION     TBD

# TROJAN

## 1.0 SITE IDENTIFICATION

Location:            Rainier, OR
License No.:         NPF-1
Docket No.:          50-344
License Status:      Permanently shutdown
Project Manager:     John Buckley

## 2.0 SITE STATUS SUMMARY

The plant was shutdown in November 1992. The DECON DP was approved in April 1996. The plant is currently undergoing dismantlement under 10 CFR 50.59. The steam generators and reactor vessel have been shipped to Hanford LLW site. The licensee was granted a site-specific Part 72 license for an onsite ISFSI in March 1999. The licensee submitted a proposed LTP in August of 1999. A license amendment approving the LTP was issued in February 2001.

The licensee began spent fuel transfer to the ISFSI in December 2002, and finished fuel transfer in August 2003. Decommissioning activities at the site are scheduled to be completed in October 2004.

## 3.0 MAJOR TECHNICAL OR REGULATORY ISSUES

None

## 4.0 ESTIMATED DATE FOR LICENSE TERMINATION    6/05

# VALLECITOS BOILING WATER REACTOR (VBWR)

## 1.0  SITE IDENTIFICATION

Location:              Sunol, CA
License No.:           DPR-1
Docket No.:            50-18
License Status:        Permanently shutdown
Project Manager:       Marvin Mendonca

## 2.0  SITE STATUS SUMMARY

The VBWR was shutdown in 1963 and NRC issued a possession only license in 1965.  The license was renewed in 1973 and the license has remained effective under the provisions of 10 CFR 50.51(b).  The facility has been maintained in SAFSTOR condition.  The site has an operating research reactor, and has hot cells that are used for power reactor fuel post irradiation examination.  The licensee plans to maintain the facility in SAFSTOR until ongoing nuclear activities are terminated and the entire site can be decommissioned.  GE has a self-guarantee instrument.  The spent fuel has been removed from the site.  There are no plans to transfer NRR project management to NMSS project management.

## 3.0  MAJOR TECHNICAL OR REGULATORY ISSUES

None

## 4.0  ESTIMATED DATE FOR LICENSE TERMINATION     TBD

# YANKEE ROWE

## 1.0  SITE IDENTIFICATION

Location:            Greenfield, MA
License No.:         DPR-3
Docket No.:          50-29
License Status:      Permanently shutdown
Project Manager:     John Hickman

## 2.0  SITE STATUS SUMMARY

The plant was permanently shut down on October 1, 1991.  The DECON DP was approved in February 1995, and the plant is undergoing dismantlement.  The steam generators were shipped to the Barnwell, North Carolina LLW facility in November 1993.  The reactor vessel was shipped to Barnwell in April 1997.  The owner has removed all of the primary systems, secondary side components, and switch yard equipment from the site.  The plant is about 80 percent dismantled.  The containment and other major structures remain.  The owner has completed construction of an onsite ISFSI.  An LTP was submitted in May 1997, and a public meeting was held to discuss the LTP in January 1998.  A public hearing was requested on the LTP but was canceled after the owner withdrew the plan in May 1999, to consider the MARSSIM approach.  The licensee resubmitted a revised LTP in November 2003 and it is currently under staff review.  The staff expects to complete its review in April 2005.  All of the fuel from the spent fuel pool has been transferred to the onsite ISFSI.  Currently the owner does not intend on terminating the license until DOE takes possession of the spent fuel in the ISFSI.

## 3.0  MAJOR TECHNICAL OR REGULATORY ISSUES

None

## 4.0  ESTIMATED DATE FOR LICENSE TERMINATION

Decommissioning complete in 2005, license termination in 2021

# ZION – UNITS 1 & 2

## 1.0 SITE IDENTIFICATION

Location:            Waukegan, IL
License No.:         DPR-39/48
Docket No.:          50-295 & 50-304
License Status:      Permanently shutdown
Project Manager:     John Hickman

## 2.0 SITE STATUS SUMMARY

Zion Units 1 and 2 were permanently shut down on February 13, 1998. The fuel was transferred to the spent fuel pool, and the owner submitted the certification of fuel transfer on March 9, 1998. A public meeting was held on June 1, 1998, to inform the public of the shutdown plans. The owner has converted the turbine-generators into synchronous condensers and have isolated the spent fuel pool within a fuel building "nuclear island." The plant has been placed in SAFSTOR, where it will remain until about 2013 when the decommissioning trust fund will be sufficient to conduct DECON activities. The owner submitted the PSDAR, site-specific cost estimate, and fuel management plan on
February 14, 2000.

## 3.0 MAJOR TECHNICAL OR REGULATORY ISSUES

None

## 4.0 ESTIMATED DATE FOR LICENSE TERMINATION     TBD

# Appendix B

## Site Summaries for
## Research and Test Reactors Undergoing
## Decommissioning

# CORNELL UNIVERSITY – TRIGA

## 1.0 SITE IDENTIFICATION

| | |
|---|---|
| Location: | Ithaca, NY |
| License No: | R-80 |
| Docket No.: | 50-157 |
| License Status: | Decommissioning Amendment |
| Project Manager: | Daniel E. Hughes |

## 2.0 SITE STATUS SUMMARY

Cornell University submitted a request for approval of a decommissioning amendment on August 22, 2003. The decommissioning of the R-80, a 500 kilowatt (kW) TRIGA reactor, will be concurrent with the decommissioning of Cornell's zero power reactor (R-89). There is no fuel on site for this reactor.

## 3.0 MAJOR TECHNICAL OR REGULATORY ISSUES

None

## 4.0 ESTIMATED DATE FOR CLOSURE     TBD

# CORNELL UNIVERSITY – ZPR

## 1.0 SITE IDENTIFICATION

Location:            Ithaca, NY
License No.:         R-89
Docket No.:          50-97
License Status:      Decommissioning Amendment
Project Manager:     Daniel E. Hughes

## 2.0 SITE STATUS SUMMARY

Cornell University submitted a request for approval of a decommissioning amendment on August 22, 2003.  The decommissioning of the R-89 (zero power) reactor will be concurrent with the R-80, a 500 kilowatt (kW) TRIGA reactor.  There is no fuel on site for this reactor.

## 3.0 MAJOR TECHNICAL OR REGULATORY ISSUES

None

## 4.0 ESTIMATED DATE FOR CLOSURE     TBD

# FORD NUCLEAR REACTOR

## 1.0  SITE IDENTIFICATION

Location:            Ann Arbor, MI
License No:          R-28
Docket No.:          50-2
License Status:      Decommissioning Amendment
Project Manager:     Patrick J. Isaac

## 2.0  SITE STATUS SUMMARY

University of Michigan submitted a request for approval of a decommissioning amendment on June 23, 2004.  There is no fuel on site for this reactor.

## 3.0  MAJOR TECHNICAL OR REGULATORY ISSUES

None

## 4.0  ESTIMATED DATE FOR CLOSURE     TBD

# GENERAL ATOMICS – TRIGA MARK F

## 1.0 SITE IDENTIFICATION

| | |
|---|---|
| Location: | San Diego, CA |
| License No: | R-67 |
| Docket No.: | 50-163 |
| License Status: | Decommissioning Amendment |
| Project Manager: | Alexander Adams |

## 2.0 SITE STATUS SUMMARY

Decommissioning activities at General Atomics (GA) are currently on hold pending the return of fuel to DOE. Fuel from GA's Mark F and Mark I reactors is currently in storage in the Mark F reactor storage pool. The licensee has dismantled the Mark F reactor to the extent possible given the storage of fuel.

## 3.0 MAJOR TECHNICAL OR REGULATORY ISSUES

DOE has refused to take the reactor fuel from GA's site unless NRC concludes that public health and safety and/or common defense and security is endangered by continuing storage at GA. The NRC staff is in the process of performing a vulnerability analysis (VA) of the GA reactors. The results of the VA will assist NRC in making a determination about the desirability of continued fuel storage at GA. DOE is concerned that accepting fuel from GA could impact legal issues surrounding DOE acceptance of fuel from the nuclear power industry.

## 4.0 ESTIMATED DATE FOR CLOSURE     TBD

# GENERAL ATOMICS – TRIGA MARK I

## 1.0 SITE IDENTIFICATION

| | |
|---|---|
| Location: | San Diego, CA |
| License No: | R-38 |
| Docket No.: | 50-89 |
| License Status: | Decommissioning Amendment |
| Project Manager: | Alexander Adams |

## 2.0 SITE STATUS SUMMARY

Decommissioning activities at GA are currently on hold pending return of fuel to DOE. Fuel from GA's Mark F and Mark I reactors is currently in storage in the Mark F reactor storage pool. The licensee has dismantled the Mark I reactor to the extent possible given the storage of fuel. To complete decommissioning activities on the Mark I reactor, the licensee needs to dismantle parts of the building in which the Mark I and Mark F reactors are located. These activities are on hold until fuel is removed from the Mark F reactor storage area.

## 3.0 MAJOR TECHNICAL OR REGULATORY ISSUES

DOE has refused to take the reactor fuel from GA's site unless NRC concludes that public health and safety or common defense and security is endangered by continuing storage at GA. The NRC staff is in the process of performing a vulnerability analysis (VA) of the GA reactors. The results of the VA will assist NRC in making a determination about the desirability of continued fuel storage at GA. DOE is concerned that accepting fuel from GA could impact legal issues surrounding DOE acceptance of fuel from the nuclear power industry.

## 4.0 ESTIMATED DATE FOR CLOSURE    TBD

# GENERAL ELECTRIC CO. – GETR

## 1.0 SITE IDENTIFICATION

| | |
|---|---|
| Location: | Sunol, CA |
| License No: | TR-1 |
| Docket No.: | 50-70 |
| License Status: | Possession Only |
| Project Manager: | Marvin M. Mendonca |

## 2.0 SITE STATUS SUMMARY

NRC issued a possession-only license for GETR on February 5, 1986. The license was renewed on September 30, 1992, to expire in 2016. The facility has been maintained in SAFSTOR condition. The site has an operating research reactor, and has hot cells that are used for power reactor fuel post irradiation examination. The licensee plans to maintain the facility in SAFSTOR until ongoing nuclear activities are terminated and the entire site can be decommissioned.

## 3.0 MAJOR TECHNICAL OR REGULATORY ISSUES

None

## 4.0 ESTIMATED DATE FOR CLOSURE    TBD

# GENERAL ELECTRIC CO. – VESR

## 1.0 SITE IDENTIFICATION

Location:            Alameda, CA
License No:          DR-10
Docket No.:          50-183
License Status:      Possession Only
Project Manager:     Marvin M. Mendonca

## 2.0 SITE STATUS SUMMARY

On April 15, 1970, NRC authorized the licensee to possess but not operate the reactor. The license was renewed on June 11, 1976, to expire in 2016. The facility has been maintained in SAFSTOR condition. The facility is next to the Vallecitos Boiling Water Reactor which is also in SAFSTOR. The licensee plans to maintain the facility in SAFSTOR until other ongoing nuclear and radioactive activities are also to be decommissioned to provide an integrated site decommission. (The site has an operating research reactor, and it also has hot cells that are used for among other things power reactor fuel post irradiation examination.)

## 3.0 MAJOR TECHNICAL OR REGULATORY ISSUES

None

## 4.0 ESTIMATED DATE FOR CLOSURE     TBD

# MANHATTAN COLLEGE

## 1.0 SITE IDENTIFICATION

Location:            Bronx, NY
License No:          R-94
Docket No.:          50-199
License Status:      Decommissioning Amendment
Project Manager:     Daniel E. Hughes

## 2.0 SITE STATUS SUMMARY

The Manhattan College license was amended on March 23, 1999, to remove authority to operate the reactor and allow possession only of the reactor.  The Amendment also approved the DP.  What remains is to complete surveys for release of the facility.  There is no fuel on site for this reactor.

## 3.0 MAJOR TECHNICAL OR REGULATORY ISSUES

None

## 4.0 ESTIMATED DATE FOR CLOSURE    TBD

**NASA – MOCKUP**

## 1.0 SITE IDENTIFICATION

Location:          NASA Glenn Research Center at Lewis Field, Cleveland, OH
License No:        R-93
Docket No.:        50-185
License Status:    Decommissioning Amendment
Project Manager:   Patrick J. Isaac

## 2.0 SITE STATUS SUMMARY

After many years of little to no activities at the Plum Brook reactor site, decommissioning is well underway. While awaiting NRC approval of the DP, NASA prepared the reactor facility by conducting pre-decommissioning activities under its POL Amendment. NRC approved the DP in March 2002. There is no fuel on site for this reactor.

## 3.0 MAJOR TECHNICAL OR REGULATORY ISSUES

None

## 4.0 ESTIMATED DATE FOR CLOSURE    2007

# NASA – PLUM BROOK

## 1.0 SITE IDENTIFICATION

Location:           NASA Glenn Research Center at Lewis Field, Cleveland, OH
License No:          TR-3
Docket No.:          50-30
License Status:      Decommissioning Amendment
Project Manager:     Patrick J. Isaac

## 2.0 SITE STATUS SUMMARY

After many years of little to no activities at the Plum Brook reactor site, decommissioning is well underway.  While awaiting NRC approval of the DP, NASA prepared the reactor facility by conducting pre-decommissioning activities under its POL Amendment.  NRC approved the DP in March 2002, and in November 2002, NASA conducted the first reactor tank entry in 30 years.  In August 2003, NASA began taking important steps in removing the reactor internals and segmenting the reactor tank for shipment to Barnwell, SC.  NASA plans to complete decommissioning by 2007.  There is no fuel on site for this reactor.

## 3.0 MAJOR TECHNICAL OR REGULATORY ISSUES

None

## 4.0 ESTIMATED DATE FOR CLOSURE     2007

# UNIVERSITY OF BUFFALO

## 1.0 SITE IDENTIFICATION

Location:            Buffalo, NY
License No:          R-77
Docket No.:          50-57
License Status:      Possession Only
Project Manager:     Daniel E. Hughes

## 2.0 SITE STATUS SUMMARY

License R-77 was amended June 6, 1997, for possession only. As of December 18, 2003, there is no firm date for DOE to accept shipment of the spent fuel. A DP has not been submitted.

## 3.0 MAJOR TECHNICAL OR REGULATORY ISSUES

As yet there is no firm shipment date for the DOE owned fuel.

## 4.0 ESTIMATED DATE FOR CLOSURE    TBD

# UNIVERSITY OF ILLINOIS

## 1.0 SITE IDENTIFICATION

Location:                    Urbana, IL
License No:                  R-111
Docket No.:                  50-151
License Status:              Decommissioning Amendment
Project Manager:             Alexander Adams

## 2.0 SITE STATUS SUMMARY

All fuel was removed from the facility during August 2004. The licensee is in the process of developing a DP for submission to NRC. There is no fuel on site for this reactor.

## 3.0 MAJOR TECHNICAL OR REGULATORY ISSUES

None

## 4.0 ESTIMATED DATE FOR CLOSURE     TBD

# UNIVERSITY OF VIRGINIA

## 1.0 SITE IDENTIFICATION

Location:            Charlottesville, VA
License No:          R-66
Docket No.:          50-62
License Status:      Decommissioning Amendment
Project Manager:     Daniel E. Hughes

## 2.0 SITE STATUS SUMMARY

The decontamination phase of decommissioning was completed as of July 2003. NRC approved the Final Status Survey Plan (FSSP), which includes the FSS for R-123 (the license for the University of Virginia's zero power pool type reactor). The staff has reviewed the Final Status Survey Report (FSSR) and an RAI has been submitted to the licensee. There is no fuel on site for this reactor.

The plan There is no fuel on site for this reactor.

## 3.0 MAJOR TECHNICAL OR REGULATORY ISSUES

None

## 4.0 ESTIMATED DATE FOR CLOSURE     TBD

# UNIVERSITY OF VIRGINIA – CAVALIER

## 1.0 SITE IDENTIFICATION

Location:         Charlottesville, VA
License No:       R-123
Docket No.:       50-396
License Status:   Decommissioning Amendment
Project Manager:  Daniel E. Hughes

## 2.0 SITE STATUS SUMMARY

The decontamination phase of decommissioning was completed as of July 2003.  NRC approved the Final Status Survey Plan (FSSP), which includes the final status survey (FSS) for R-66 (the license for the University of Virginia's 2MW pool type reactor).  The staff has reviewed the Final Status Survey Report (FSSR) and an RAI has been submitted to the licensee.  There is no fuel on site for this reactor.

## 3.0 MAJOR TECHNICAL OR REGULATORY ISSUES

None

## 4.0 ESTIMATED DATE FOR CLOSURE     TBD

# UNIVERSITY OF WASHINGTON

## 1.0  SITE IDENTIFICATION

Location:             Seattle, WA
License No:           R-73
Docket No.:           50-139
License Status:       Decommissioning Order
Project Manager:      Alexander Adams

## 2.0  SITE STATUS SUMMARY

Little progress had been made towards decommissioning because the licensee could not acquire decommissioning funding from the State of Washington as a separate budget item. Because of this, the University of Washington has decided to fund decommissioning through local University funds and is now moving forward.  Contractors have been put in place to prepare for decommissioning activities.  It is anticipated that decommissioning activities will start Winter 2005.  There is no fuel on site for this reactor.

## 3.0  MAJOR TECHNICAL OR REGULATORY ISSUES

None

## 4.0  ESTIMATED DATE FOR CLOSURE     TBD

**VETERANS ADMINISTRATION**

1.0  SITE IDENTIFICATION

Location:            Omaha, NE
License No:          R-57
Docket No.:          50-131
License Status:      Operating License
Project Manager:     Alexander Adams

2.0  SITE STATUS SUMMARY

The licensee has submitted a DP that is being reviewed by NRC for acceptance.  There is no fuel on site for this reactor.

3.0  MAJOR TECHNICAL OR REGULATORY ISSUES

None

4.0  ESTIMATED DATE FOR CLOSURE     TBD

# WESTINGHOUSE

## 1.0 SITE IDENTIFICATION

Location:            Waltz Mill Site, New Stanton, PA
License No:          TR-2
Docket No.:          50-22
License Status:      Decommissioning Amendment
Project Manager:     Patrick J. Isaac

## 2.0 SITE STATUS SUMMARY

The TR-2 License was amended in March 1963 to allow possession, but not use of the reactor. The Westinghouse Test Reactor, located on the Waltz Mill site, is undergoing decommissioning in accordance with the Final DP which was approved in September 1998. CBS (formerly Westinghouse Electric Corporation), which operated the Waltz Mill Facility, was the licensee of the TR-2 and SNM-770.

Radiological contamination in soil and groundwater exist on a portion of the site as a result of the clean-up activities following a 1961 incident at the test reactor, waste segregation activities, and nuclear laundry services. Significant contamination is also present in retired facilities (hot cells, hot cell support rooms, and a section of the fuel transfer canal) within one of the site buildings. Contaminants are primarily strontium-90 (Sr-90) and cesium-137 (Cs-137), with lesser quantities of mixed fission, activation products, and trace levels of transuranic radionuclides. The TR-2 DP required removal of designated portions of the shutdown reactor as necessary and sufficient to terminate the Part 50 portion of the license. At that point, the remaining residual radioactive materials would be transferred to SNM-770 where they would continue to be controlled under that license. The DP did not include or provide for any criteria or provide for any unrestricted release of the facility. In March 1999, Viacom acquired the TR-2 license and a new company, Westinghouse Electric Company, LLC, (Westinghouse) became the holder of the SNM-770 License. Westinghouse and Viacom entered into a project management agreement whereby Westinghouse agreed to act as Viacom's decommissioning project manager for the TR-2 reactor. The pressure vessel and pressure vessel internals have been removed in accordance with the DP, as has all of the biological shield that needed to be removed in order to remove the pressure vessel. Two provisions of the DP still need to be accomplished: determining the residual radioactivity remaining in situ and preparing the necessary amendments for and requesting the transfer to the SNM-770 license. There is no fuel on site for this reactor.

## 3.0 MAJOR TECHNICAL OR REGULATORY ISSUES

Westinghouse is refusing to accept the transfer to the SNM-770 license. Viacom filed a 10 CFR 2.206 petition in which it alleges that Westinghouse is in violation of 10 CFR 50.5, Deliberate Misconduct. Westinghouse claims that Viacom did not perform all the actions required prior to the transfer. NRC issued a Director's Decision concluding that Westinghouse was not in violation of 10 CFR 50.5 on August 26, 2003. Viacom and Westinghouse are currently engaged in a commercial dispute and are under arbitration to resolve the disputed issues.

## 4.0 ESTIMATED DATE FOR CLOSURE    2005

# Appendix C

## Site Summaries for
## Current Complex Sites Undergoing
## Decommissioning

# AAR MANUFACTURING, INC.

## 1.0 SITE IDENTIFICATION

| | |
|---|---|
| Location: | Livonia, MI |
| License No.: | STB-0362 |
| Docket No.: | 04000235 |
| License Status: | Terminated |
| Project Manager: | Kristina Banovac |

## 2.0 SITE STATUS SUMMARY

Thorium (Th) contaminated surface and subsurface soil has been identified at several locations in open land areas on the site. Ground water contamination is not present.

AAR submitted the final remediation plan (RP) on October 14, 1997, and NRC approved the RP on May 22, 1998. Remediation at the site began on October 12, 1998. Geoprobe sampling identified additional soil contamination on the western side of the property.

On September 17, 1999, AAR submitted a proposed revision to the approved RP. The proposed plan involved remediation of only soils containing thorium concentrations exceeding 116 pCi/g, which is the unimportant quantity (0.05 weight percent) of source material, exempt from regulation, established in 10 CFR 40.13(a). After staff consultation with the Commission on this policy issue, NRC informed AAR that the revised remediation approach was not acceptable, by letter dated August 9, 2002.

The cost of decommissioning is unknown at this time.

## 3.0 MAJOR TECHNICAL OR REGULATORY ISSUES

AAR is not a licensee. This site was owned and operated by Brooks & Perkins, Inc. from 1959 until the license was terminated in 1971. AAR purchased Brooks & Perkins in 1981. Since AAR is not directly responsible for the contamination onsite, it believes it should not be responsible for the cost of site remediation. If remediation costs become large, it is possible that AAR may legally challenge its responsibility to fund the remediation activities.

AAR is currently pursuing the restricted release option for a portion of its site and plans to enter into a settlement agreement with NRC on the restrictions and controls needed for restricted release. The agreement would include using a deed restriction that would outline the restrictions on the site, such as prohibiting farming and developing residential properties on the site; the deed restriction would transfer to each subsequent owner of the property through the deed. The agreement and deed restriction would allow NRC or local and State governments to monitor and enforce the restrictions. Once AAR submits its restricted release DPs, the staff will complete its review and inform the Commission of its results and any policy issues that result from AAR's proposal. The staff is currently working with AAR to resolve the technical issues at the site.

Elevated levels of thorium have also been identified along the fence separating AAR and CSX Transportation, Inc. (CSX). Although contamination appears to be very limited, there is the potential that financial responsibility for the contamination on CSX property may become an issue. No remediation has been performed by CSX.

To date, public interest in remediation activities at the site is minimal.

4.0 ESTIMATED DATE FOR CLOSURE     1/07

# ABB PROSPECTS, INC. (FORMERLY C.E. WINDSOR)

## 1.0 SITE IDENTIFICATION

Location:          Windsor, CT
License No.:       06-00217-06
Docket No.:        030-03754
License Status:    Timely Renewal
Project Manager:   Laurie Kauffman, R I

## 2.0 SITE STATUS SUMMARY

The ABB Prospects, Inc., (formerly Combustion Engineering-Windsor) site consists of soils and building and equipment surfaces contaminated with uranium and by-product material from operations that occurred from the late 1950s until 2001.

A revised site-wide DP was received by NRC on April 7, 2003. Estimated time for remediation of areas associated with NRC licensed activities was approximately one year after approval of the DP. The acceptance review of the DP was delayed until the licensee submitted a revised dose modeling scenario including revised DCGLs. The revised DCGLs were submitted September 2003. On June 1, 2004, the license was amended to incorporate the DP.

The licensee submitted an amendment request on April 7, 2004 for an increase in the U-235 possession limits and for an exemption from criticality monitoring requirements specified in 10 CFR 70.24(a). A RAI was forwarded to the licensee, and Region I is coordinating with the licensee and NMSS to obtain the additional information.

Under the current license, the licensee is removing interior systems, components, ducts, piping, conduits, and so forth from the buildings in Building Complexes 2, 5, and 17. Equipment and material are being released from the site using Regulatory Guide 1.86 criteria as permitted by the current license. The present license also permits the licensee to demolish the buildings of Building Complexes 2, 5, and 17 down to grade level only.

The licensee estimates the cost of decommissioning to be approximately $2.6 million, based on the licensee's Decommissioning Funding Plan dated December 2003.

## 3.0 MAJOR TECHNICAL OR REGULATORY ISSUES

Although the State of Connecticut had filed a hearing request, which was later withdrawn, public interest in the area is not high.

## 4.0 ESTIMATED DATE FOR CLOSURE     12/07

# ALLIANT ORDINANCE AND GROUND SYSTEMS, LLC (ATK)

## 1.0 SITE IDENTIFICATION

Location:           Twin Cities Army Ammunition Plant, Arden Hills, Minnesota
License No.:        SUB-00971
Docket No.:         040-07982
License Status:     Decommissioning
Project Manager:    George M. McCann, RIII

## 2.0 SITE STATUS SUMMARY

The Twin Cities Army Ammunition Plant, in Arden Hills, Minnesota, was used to produce munitions and ordnance materials containing depleted uranium (DU) for the U.S. Military. Contamination at the site consists mainly of DU. In addition, NRC inspectors identified a small area next to the parking lot containing Cs-137, which was unexpected since ATK's license had not authorized the possession of Cs-137. The residual Cs-137 contamination identified in the soil is below NRC's screening value of 11 pCi/g, as cited in NUREG-1757. The inspectors also determined that a former NRC licensee, Minnesota, Mining and Manufacturing [22-00057-06 (terminated in 1994)] had processed unsealed Cs-137 in facilities near the parking lot, and the low levels of residual Cs-137 is most likely from these past operations.

ATK submitted its DP in October 1997 and it was approved in June 1998. The licensee submitted an addendum to the DP in June 2001, which was approved in October 2001. The licensee immediately started decommissioning of its facilities in October 2001, and completed demolition and remediation of the former production facilities and underlying soils and sewers, early in 2003. All decommissioning waste have been shipped off site, and these surveys constitute the last onsite phase of ATK's decommissioning activities. The licensee submitted its license termination and FSSR on June 18, 2004. The staff anticipates completing the review of this report during September 2004.

The licensee estimates the cost of decommissioning to be approximately $6.6 million.

## 3.0 MAJOR TECHNICAL AND REGULATORY ISSUES

None

## 4.0 ESTIMATED DATE FOR CLOSURE    12/04

# AUGUSTANA COLLEGE

## 1.0 SITE IDENTIFICATION

Location:          Sioux Falls, SD
License No:        40-06921-03
Docket No:         030-01063
License Status:    Active
Project Manager:   Robert Evans, R IV

## 2.0 SITE STATUS SUMMARY

The licensee has a small research/ teaching program confined to the college with no temporary job sites. The licensee uses microcurie quantities of mostly P-32, S-35, and I-125 in unsealed form in research laboratories. In addition, the College possesses numerous microcurie quantity sealed sources used for classroom demonstrations and teaching experiments.

On February 17, 2003, the licensee requested the decommissioning approval of its former 10 CFR 20.304 burial site located on campus property. The burial site apparently consisted of C-14 contaminated waste. The maximum amount of C-14 buried was assumed to be the total amount of C-14 that was ordered, 12 mCi. The licensee believes that only experimental waste materials were buried, therefore, the actual amount buried was probably much less than 12 mCi. The licensee requested that the burial site be free released without being remediated.

A site assessment was performed by the licensee using the NRC-approved RESRAD computer code. The RESRAD program, using default parameters, determined that the site met the unrestricted use dose limit of 25 mRem/yr. The RESRAD summary submitted by the licensee calculated the dose in 1969 to be 77.8 mRem/yr. At year 30 (1999), the program shows a dose reduction to 0.00 mRem/yr.

Preliminary review of the decommissioning activities began during the 3rd quarter of 2003. A letter dated July 15, 2003, was sent to the licensee requesting additional pertinent information necessary for an environmental assessment review. FCDB received the additional information from the licensee and prepared an EA which is currently under review by NRC (NMSS/DWMEP) for comment and concurrence.

The cost of decommissioning should be limited to administrative costs. The licensee has requested NRC authorization to free release the property without having to remediate the former burial site.

## 3.0 MAJOR TECHNICAL OR REGULATORY ISSUES

None

## 4.0 ESTIMATED DATE FOR CLOSURE     9/04

# BABCOCK & WILCOX (SHALLOW LAND DISPOSAL AREA)

## 1.0  SITE IDENTIFICATION

Location:            Parks Township, Armstrong County, PA
License No.:         SNM-2001
Docket No.:          07003085
License Status:      Active
Project Manager:     Amir Kouhestani

## 2.0  SITE STATUS SUMMARY

The site consists of 10 trenches that were used to dispose of wastes, scrap, and trash from a nearby nuclear fuel fabrication facility in Apollo, PA.  Principal radioactive contaminants at the site are natural, enriched, and DU, and lesser quantities of Am-241, plutonium, and thorium.

This site is designated by USACE as a Formerly Utilized Sites Remedial Action Program (FUSRAP) site.  In December 2001, Congress directed USACE to remediate the site.  In March 2002, USACE issued a final site Preliminary Assessment (PA) in accordance with the Comprehensive Environmental Response, Compensation, and Liability Act of 1980, as amended (CERCLA).  The PA concludes that USACE will remediate the site in accordance with CERCLA and FUSRAP requirements, and consistent with the USACE-NRC Memorandum of Understanding (MOU).  In December 2001, staff conditioned the B&W Parks Shallow Land Disposal Area (BWXT–SLDA) license to allow for an eventual suspension of the license when USACE completes its ROD for the site under CERCLA.  In June 2003, USACE initiated its onsite investigative activities.  USACE has discussed an accelerated 2005 schedule for issuance of its ROD.

No financial assurance issues have been identified at this time.  The cost of decommissioning is estimated to be approximately $10 million.

## 3.0  MAJOR TECHNICAL OR REGULATORY ISSUES

In the event that USACE's congressionally mandated site remediation does not take place, NRC staff anticipates that BWXT–SLDA may request license termination, with restrictions on future land use.  The PADEP has stated that it will not assume responsibility for the site (i.e., become the institutional control authority) if it is decommissioned with land-use restrictions.

There is significant public and Congressional interest in the site.  Congressman Murtha is closely following USACE's site remediation efforts.

No financial assurance issues have been identified at this time.

## 4.0  ESTIMATED DATE OF CLOSURE        10/09

# BATTELLE COLUMBUS LABORATORIES

## 1.0 SITE IDENTIFICATION

| | |
|---|---|
| Location: | West Jefferson, OH |
| License No: | SNM-00007 |
| Docket No: | 070-00008 |
| License Status: | Decommissioning only |
| Project Manager: | Mike McCann, R III |

## 2.0 SITE STATUS SUMMARY

Battelle Memorial Institute (BMI) performed atomic energy research and development (R&D) for the DOE and its predecessor agencies between 1943 and 1986 at its Columbus Laboratories sites. The nuclear research included fabrication of uranium fuel elements; reactor development; submarine propulsion research; fuel reprocessing; and safety studies of reactor vessels and piping.

A total of six buildings are located at the Battelle research (Nuclear Sciences Area ) location near West Jefferson, Ohio. The Nuclear Sciences Area occupies an 11-acre fenced enclosure in the northern portion of the West Jefferson site. This enclosed facility consists of four major buildings, a guardhouse, and several smaller structures on a bluff overlooking Darby Creek and Battelle Lake. Three of the major buildings and their support structures are the focus of the final phase of the decommissioning project. Outside of the fenced area, an abandoned filter bed and part of the site sanitary sewer system are also included in the project. There are no indications that significant radiological impacts on sub-surface soils or ground water exist.

The licensee estimates the cost of decommissioning to be approximately $247 million.

## 3.0 MAJOR TECHNICAL OR REGULATORY ISSUES

Battelle is the licensee for the project, but based on previous agreements with DOE, is not funding the majority of the cleanup costs. The DOE, as the primary party responsible for funding the clean-up of the site, has made a business decision to discontinue using Battelle as the contractor for performing the direct decontamination and remediation activities at the Battelle West Jefferson site. Instead, DOE will use an independent contractor reporting directly to DOE. Funding for the project is expected to be adequate through the final phase of cleanup activities. However, the contractor will conduct the work under the authority of the BMI license, and BMI will continue to be responsible for site clean-up and over-sight of the contractor. NRC will continue to hold BMI accountable for meeting the cleanup schedule in accordance with all applicable decommissioning standards and regulations.

## 4.0 ESTIMATED DATE FOR CLOSURE     12/05

# CABOT PERFORMANCE MATERIALS, INC.

## 1.0 SITE IDENTIFICATION

Location:            Reading, PA
License No.:         SMC-1562
Docket No.:          04009027
License Status:      Active
Project Manager:     Ted Smith

## 2.0 SITE STATUS SUMMARY

Contamination at the site consists of surface and subsurface uranium and thorium contamination, in the form of slag. Ground water contamination has not been identified at the site.

The March 2000 DP, as supplemented in November 2002, proposes unrestricted release of the site in its current condition. NRC staff issued a RAI in March 2003, for additional information regarding site characterization, source term modeling, and previously unconsidered aspects of meeting the "as low as is reasonably achievable" requirements of the LTR at the site. The licensee provided a proposed conceptual approach to resolving NRC questions in Spring 2004. The conceptual approach includes emplacement of a riprap cover as an engineered barrier.

The licensee will be reevaluating its cost estimate as part of the revised DP incorporating the rip-rap cover design.

## 3.0 MAJOR TECHNICAL OR REGULATORY ISSUES

While conducting research and analysis on slags, the NRC staff identified potential issues regarding both the quantity and concentration of radioactive slag at the site. These and similar questions have been raised by PADEP as significant concerns about the adequacy of the characterization of the site. By letter dated August 27, 2004, the staff agreed with the concept of the riprap cover. The licensee will submit a revised DP implementing the approach.

No major financial assurance issues are associated with this site. A potential financial assurance concern would arise if off-site disposal were required.

Public interest in the decommissioning activities at the site has been increasing since late 2002.

## 4.0 ESTIMATED DATE FOR CLOSURE     9/05

# CURTISS-WRIGHT CHESWICK

## 1.0 SITE IDENTIFICATION

Location:           Cheswick, PA
License Nos.:        SNM-1120
Docket Nos.:         070-01143,
License Status:      Active
Project Manager:     Mark Roberts, R I

## 2.0 SITE STATUS SUMMARY

The Curtiss-Wright Electro-Mechanical Corporation (formerly Westinghouse Government Services) facility is a multi-building complex situated on 110 acres near Cheswick, PA. The Allegheny River is approximately one mile south of the facility. Past commercial and government fuel processing operations with low-enriched and high-enriched uranium have left a legacy of contamination in, on, around, and under some of the site buildings. Buildings 4, 5, 5A, 5B, 5C, and 5F show some fixed contamination in generally inaccessible areas or contamination in drain lines. Fuel processing activities ceased many years ago; however, the remaining contamination in many locations is deposited in areas that are inherent to the design and structure of the facility (i.e., exterior load-bearing walls, structural steel supports, drain lines beneath significant equipment, and roof supports). The facility has two other NRC radioactive materials licenses for contaminated motor servicing and radiography. Buildings 4 and 5 are presently being used for manufacturing and support functions for its pump and motor operations. Contaminated debris was uncovered in 1984 in a former ballfield, a 2-acre area south of the main buildings. This area requires further evaluation to characterize the radiological conditions.

Remediation at the facility has been ongoing using criteria identified in the license. A comprehensive RP for addressing the remaining contaminated areas of the site is under development, but has not yet been submitted to NRC. In general, remediation work was scheduled so that the activities did not interfere with the manufacturing operations. Surveys for the remediated areas have been performed by the licensee, but final survey documentation has not been transmitted to NRC.

The licensee's decommissioning funding plan provides $750,000 for financial assurance. Adjustments to this amount may be required after site characterization is completed.

## 3.0 MAJOR TECHNICAL OR REGULATORY ISSUES

Although a significant amount of decontamination work has been performed in both interior and exterior areas of the facility, the licensee has not completed a FSS report for the remediated area.

Monitoring well data for identification of any groundwater issues is limited.

PADEP has interest in the site, particularly in the suspect ballfield area.

## 4.0 ESTIMATED DATE FOR CLOSURE     12/08

# DEPARTMENT OF THE ARMY

## 1.0 SITE IDENTIFICATION

Location:              Fort McClellan, AL
License No.:           01-02861-05
Docket No.:            030-17584
License Status:        Terminated
Project Manager:       Orysia Masnyk Bailey, R II

## 2.0 SITE STATUS SUMMARY

This site was licensed from 1956 until 1973 and from 1980 until the present. Building 3192 housed a hot cell for the fabrication of cesium and cobalt sources and as a result, building surfaces, soil and below ground tanks became contaminated. A license (01-02861-04) was issued for the possession of this residual contamination in Buildings 3182 and 3192 in 1974. Starting in 1980, the licensee performed closeout surveys for most of its authorized places of use under the Broad Scope license. Buildings 3182 and 3192 were remediated under a DP dated March 31, 1995. License No. 01-02861-04 was terminated on October 19, 1998, following an NRC confirmatory survey.

Under the Broad Scope license, the Army performed closeout surveys and NRC performed confirmatory surveys of several areas including the burial areas at Iron Mountain and Rattlesnake Gulch, and of Buildings 1081, 2281, 3180, 3181, 3182, 3185, T-810, T-811, T812, T-836, T-837 and Alpha Field. Based on a characterization survey that caused the licensee to suspect the presence of discrete sources, a DP dated March 2, 1999, was submitted for the remediation of the Burial Mound at Pelham Range. The mound was found to include discrete sources (Co-60) and some Sr-90. The licensee provided a closeout survey. An NRC confirmatory survey was performed July 5-6, 2003. NRC and licensee survey results disclosed no contamination significantly above background levels. Except for the Pelham Range, the remainder of the Fort McClellan property has been turned over to the State of Alabama.

The staff is not aware of a specific estimate for the cost of decommissioning.

## 3.0 MAJOR TECHNICAL OR REGULATORY ISSUES

As part of the final site closeout survey, the Army performed a fly over survey of the base which resulted in the discovery of an additional burial ground. Elevated readings were seen at the Anniston City Recreational Area (LaGarde Park), a property that the Army donated to the city in 1976. Ground investigation disclosed the presence of soil contaminated with cesium and cobalt. USACE took responsibility for the site. In August 2003, USACE initiated clean up action at this site under CERCLA. Under this cleanup, contaminated soil was identified and removed as funds allowed. USACE returned to the site in 2004 to perform a characterization survey to gather data for a risk analysis regarding the Pelham Range burial area. The license will be terminated when ACE remediates the remaining burial area. The State of Alabama and the EPA are monitoring clean up activities.

## 4.0 ESTIMATED DATE FOR CLOSURE    6/05

**DOW CHEMICAL COMPANY (DOW)**

1.0 SITE IDENTIFICATION

Location:             Bay City, MI
License No.:          STB-527
Docket No.:           04000017
License Status:       Active
Project Manager:      David Nelson

2.0 SITE STATUS SUMMARY

Contamination at the site consists of thorium contaminated slag storage piles. Ground water contamination at the site has been identified.

Dow submitted a DP on October 12, 1995. Dow is requesting unrestricted release of the site. The DP was approved in July 1997. In September 2000, Dow informed NRC that decommissioning of the Bay City site had been complicated by a larger volume of contamination than originally estimated, the presence of wetlands, and winter flooding. In December 2003, Dow submitted a revised supplement to amend the previously approved DP. The staff is currently reviewing the DP.

There are no immediate radiological hazards at the site.

The licensee has not yet submitted an estimate of the cost of decommissioning. Dow will submit a decommissioning funding plan with a detailed cost estimate upon acceptance of a decommissioning approach, but before approval of the DP.

3.0 MAJOR TECHNICAL OR REGULATORY ISSUES

The staff has not identified any major off-site environmental issues that will not be addressed during decommissioning of the facility.

There has been minimal public interest in the decommissioning activities at this facility.

4.0 ESTIMATED DATE FOR CLOSURE     4/06

# EGLIN AIR FORCE BASE

## 1.0 SITE IDENTIFICATION

| | |
|---|---|
| Location: | Walton County, FL |
| License No.: | 42-23539-01AF |
| Docket No.: | 030-28641 |
| License Status.: | Active |
| Project Manager: | Robert Evans, R IV |

## 2.0 SITE STATUS SUMMARY

The Department of the Air Force submitted a DP to NRC on May 24, 2002. The licensee subsequently submitted supplemental information dated November 1, 2002, and August 21, 2003. The licensee requested that the DP for Test Area C-74L at Eglin AFB be approved. NRC is considering the issuance of an amendment to Materials License 42-23539-01AF to approve the DP.

Test Area C-74L is located in Walton County, Florida, within the north-central portion of Eglin AFB. From 1974 to 1978, the area was used for pre-production testing of a gun system which used depleted uranium ammunition. An estimated 16,315 pounds of depleted uranium were expended at the site. The test area currently consists of a 4-acre radiologically controlled area, fire control/ballistics building, gun corridor, target area, well house building, drum storage area, and surrounding land.

The decommissioning cost estimate provided in the DP was approximately $540,000 (excluding disposal costs) plus an additional $37,000 for the final status survey.

## 3.0 MAJOR TECHNICAL OR REGULATORY ISSUES

The licensee commenced with reclamation activities at risk under the authority of an existing Air Force permit. The licensee chose to conduct decommissioning early because funding was available only for a certain time frame and because the work can only be conducted during seasonal breaks in munitions testing.

The licensee proposed a derived concentration guideline level (DCGL) of 600 pCi/g for uranium in soil. Based on detailed dose modeling NRC proposed, and the licensee accepted, a DCGL of 469 pCi/g. The NRC-approved DCGL exceeds the value listed in the NRC–EPA Memorandum of Understanding; therefore, a Level 1 EPA consultation is necessary for this site.

The licensee recently stated that it had unexpectedly discovered depleted uranium contamination below the ground surface. Further, the Air Force was experiencing funding problems for this project. The reclamation work is expected to be completed during the first quarter of fiscal year 2005.

## 4.0 ESTIMATED DATE FOR CLOSURE     TBD

# ENGELHARD MINERALS – ILLINOIS

## 1.0 SITE IDENTIFICATION

Location:          Great Lakes Naval Training Center, Great Lakes, IL
License No:        SMC-01207, SUC-01332
Docket No:         040-08306, 040-08680
License Status:    Terminated
Project Manager:   Eugenio Bonano, R III

## 2.0 SITE STATUS SUMMARY

Engelhard Minerals & Chemicals Corporation (Engelhard), which is no longer in business, was licensed to repackage and ship monazite sand from the Great Lakes Naval Training Center to other U.S. Atomic Energy Commission (AEC)/NRC licensees. The area was used by the U.S. General Services Administration (GSA), which transferred control to the Defense Logistics Agency. The Engelhard license to ship the material was terminated in 1975 (SMC-01207), and 1983 (SUC-01332). The former licensee was authorized to possess 119,829.33 kilograms (SMC-01207) and 67,965 kilograms (SUC-01332) of natural thorium (Monazite Sand). The Navy, which is the site owner, assumed responsibility for the Great Lakes site cleanup.

A scoping survey conducted in March 2000, indicated radiological concentrations of Th-232 ranging from 0.93 picocuries per gram (pCi/g) to 64.31 pCi/g with an average concentration of approximately 17.0 pCi/g. The monazite sand encompasses an area of approximately 90,000 square yards (yds) in a former tank farm area located within the boundaries of the Great Lakes Naval Training Center. Due to the relatively insoluble nature of the thorium, groundwater impact is not a concern.

Characterization, cleanup, and FSSs are being done in three phases. In Phase 1, the site formerly known as Tank farm #5, was characterized. The entire tank farm was surveyed and surface soil samples were collected and analyzed for Thorium-232. The remainder of the Tank farm was fenced to restrict access pending further investigation and remediation. In Phase II, an excavated soil pile area was remediated and the contaminated soil was shipped for disposal. The third phase involves additional remediation and a FSS of the north fence area. Characterization samples will be collected and analyzed on-site, followed by remediation and FSS.

The licensee estimates the cost of decommissioning to be approximately $1.3 million.

## 3.0 MAJOR TECHNICAL OR REGULATORY ISSUES

This is an unlicensed facility. However, the Navy has assumed responsibility for the clean-up of this formerly licensed site. The Navy has been working cooperatively with the NRC Region III staff, and has agreed to employ NRC regulations and guidance documents, such as MARSSIM and NUREG-1757, to clean-up the site.

## 4.0 ESTIMATED DATE FOR CLOSURE:    12/05

# ENGELHARD MINERALS – OHIO

## 1.0 SITE IDENTIFICATION

Location:           Ravenna Army Ammunition Depot (RVAAP), Ravenna, OH
License No:         SMC-01207
Docket No:          040-08306
License Status:     Terminated
Project Manager:    Eugenio Bonano

## 2.0 SITE STATUS SUMMARY

The Engelhard Minerals & Chemicals Corporation (Engelhard), which is no longer in business, was licensed to repackage and ship monazite sand from the Ravenna Army Ammunition Plant to other AEC/NRC licensees. The area was used by the GSA, which later transferred control to the Defense Logistics Agency. The Engelhard license was terminated in 1975. The Army is the site owner and assumed responsibility for the site cleanup of monazite sand contamination. The Th-232 activity concentration ranged from a low of 7 pCi/g to a high of 1,650 pCi/g. Due to the relatively insoluble nature of the thorium, groundwater impact is not a concern.

The Army submitted an amended FSS work plan documenting a DCGL value of 4.0 pCi/g for the release criteria. The NRC reviewed and concurred on the plan. The Army began work in July 2004 and expects to complete sample analysis and to issue an FSSR by December 2004.

The licensee estimates the cost of decommissioning to be approximately $500,000.

## 3.0 MAJOR TECHNICAL OR REGULATORY ISSUES

This is an unlicensed facility. However, the Army has assumed responsibility for the clean-up of this formerly licensed site. The Army has been working cooperatively with the NRC Region III staff and has agreed to employ NRC regulations and guidance documents, such as MARSSIM and NUREG-1757, to clean-up the site.

## 4.0 ESTIMATED DATE FOR CLOSURE:    3/05

# FMRI, INC.  (FORMERLY FANSTEEL)

## 1.0  SITE IDENTIFICATION

| | |
|---|---|
| Location: | Muskogee, OK |
| License No.: | SMB-911 |
| Docket No.: | 040-07580 |
| License Status: | Expired (possession only) |
| Project Manager: | Jim Shepherd |

## 2.0  SITE STATUS SUMMARY

Contaminants at the site include natural uranium and decay products, and natural thorium and decay products.  Chemical contamination in the form of metals including tantulum, niobium, chromium, antimony, tin, barium, arsenic; ammonia fluoride and methyl isobutyl ketone are also present.  Soil and groundwater contamination are non-uniformly distributed.

Fansteel decontaminated approximately 35 acres of the 110-acre Muskogee facility designated as the "Northwest Property," and NRC released this area for unrestricted use.  Fansteel had an NRC license dated March 25, 1997, to complete the processing of ore residues, calcium fluoride residues, and wastewater treatment residues containing uranium and thorium, in various site impoundments.  The current license expired in September 2002; the renewal application was denied because Fansteel wrote off the cost of the facility in its bankruptcy and did not provide sufficient financial assurance.

In November 2001, Fansteel suspended all operations at the Muskogee site, and in January 2002, filed Chapter 11.  Subsequently, NRC drew on the financial assurance instruments, and that money is now in a standby trust.

On November 17, the bankruptcy court approved Fansteel's reorganization plan, to divide the company into two parts, with one part going to the other creditors.  FMRI, a new subsidiary of Reorganized Fansteel, would become the licensee for the Muskogee site.  On December 4, 2003, NRC approved the request for exemption to financial assurance requirements, the license transfer and the DP, subject to the bankruptcy reorganization plan becoming effective.  The approved DP outlines remediation that focuses on the most risk-significant areas and accomplishes those activities first.  The approval also authorizes FMRI to draw up to $2 million from the standby trust for remediation work if it has insufficient funds from Fansteel to continue the work.  The reorganization plan became effective January 23, 2004.

NRC granted a request for a hearing by the Oklahoma Attorney General on the DP on November 3 (LBP-03-22).  On May 26, 2004, the ASLB issued a decision to "uphold the NRC staff's issuance of the license amendment in question" (LBP-04-08).  Following agreement by FMRI to the State to conduct chemical characterization of the site, the State did not appeal the ASLB ruling.

The licensee estimates the cost of decommissioning to be approximately $40 million; in the bankruptcy filing, it estimated the cost to be $57 million.

## 3.0 MAJOR TECHNICAL OR REGULATORY ISSUES

Fansteel has provided a total of about $4.5 million in financial assurance. The original estimate, by deposition to the Bankruptcy Court, that Fansteel provided to decommission the site was $57 million; this was for off-site disposal of all wastes greater than 10 pCi/g total, that is a license condition limit. The revised estimate of $26 million is based on dose criteria of 10 CFR 20.1402 using an industrial land use scenario with no ground water pathway. It estimated an additional $14 million for commitments to Oklahoma Department of Environmental Quality (ODEQ), primarily ground water remediation. Because it is in a bankruptcy proceeding, Fansteel stated it is not able to provide the additional financial assurance.

FMRI has stated it will not commence remediation by September 1, 2004, as previously committed to in the letter of May 8, 2003. This letter formed a significant part of the basis on which NRC approved the DP, which was necessary for Fansteel to file a bankruptcy reorganization plan. Staff is currently working with Fansteel to address this issue.

There is high public interest from the State of Oklahoma and the Cherokee Nation.

## 4.0 ESTIMATED DATE FOR CLOSURE    2023+

# HERITAGE MINERALS, INC.

## 1.0  SITE IDENTIFICATION

Location:             Lakehurst, NJ
License No.:          SMB-1541
Docket No.:           040-08980
License status:       Renewed – 9/20/99
Project Manager:      Craig Gordon, R I

## 2.0  SITE STATUS SUMMARY

Contamination at the site consists of monazite sand from process operations involving rare mineral extraction.

Heritage Minerals, Inc., (HMI) submitted its DP/FSSP in November 1997.  NRC approved the DP/FSSP in October 1999.  The HMI DP/FSSP provided the basis for disposal of thorium contaminated sand and remediation of mill buildings and equipment.  HMI requested unrestricted release for the site after license termination.

HMI did not meet the 24-month requirement to complete site decommissioning, and as a result, a predecisional enforcement conference was held on January 8, 2003, to discuss a potential violation.  The decision to take enforcement action is being held in abeyance pending HMI' s commitment to resolving those issues related to site remediation of licensable material found during NRC confirmatory surveys.  Following the conference, the licensee undertook substantial additional remediation a the site.  This included demolition of the mill buildings and cleanup and release of equipment.  The licensee submitted a proposal to resolve all remaining issues on June 30, 2004.  The proposal is currently under NRC review.

The licensee's financial assurance to cover the cost of decommissioning is by Letter of Credit in the amount of $400,000.

## 3.0  MAJOR TECHNICAL OR REGULATORY ISSUES

HMI does not believe they should be responsible for contaminated areas from previous operations which are below the level for exempt source material quantities.  The entire site covers a large area in Lakehurst, New Jersey, while the licensed material was limited to a very small outdoor area and mill buildings.  NRC-licensed portions of the site are within an area of enhanced background and varying concentrations of source material, some above licensable levels.  This has raised regulatory issues with New Jersey over final cleanup and continued radiological exposure if NRC terminates the license.  HMI has requested NRC approval of an amendment to define a boundary for NRC licensed material.  However, the State believes that NRC jurisdiction should extend to other areas which contain exempt quantities of uranium and thorium, but do not exceed unrestricted use criteria.  The primary State issue is that once NRC terminates the license, the large contaminated areas of the site not covered by the license could involve an extended and costly remediation in order to meet the State's unrestricted release criteria.

## 4.0  ESTIMATED DATE FOR CLOSURE     6/05

# HOMER LAUGHLIN

## 1.0 SITE IDENTIFICATION

Location:            Newell, WV
License No.:         SUB-00081
Docket No.:          040-01957
License Status:      Terminated
Project Manager:     Craig Gordon, R I

## 2.0 SITE STATUS SUMMARY

The Homer Laughlin China Company (HLC) is a formerly licensed site. HLC was licensed by the AEC for possession of 100,000 pounds of source material used as a glazing agent (up to 20% uranium) in the production of ceramic tableware. The license was terminated in 1972 based upon a letter from HLC stating that all remaining licensed materials had been returned to their supplier. A review of the terminated license file determined that no record of licensee closeout survey or NRC confirmatory survey was performed.

In 1994, approximately 500 pounds of depleted uranium oxide ($U_3O_8$) sand was discovered on the property. A contractor was hired to survey areas where licensed materials were used and stored and provide a radiological characterization of material in the facility. Several areas of fixed and removable contamination exceeding NRC limits for unrestricted use were identified during the characterization survey. A Confirmatory Action Letter issued to HLC outlined HLC's commitment to package and dispose of the bulk source material, limit access to contaminated areas, and submit a DP. After NRC approved the DP in January 1995, HLC and its contractor initiated facility decommissioning.

HLC did not complete decommissioning in production areas because they were unable to remove fixed contamination from surfaces of equipment and structures which exceeded NRC unrestricted release guidelines using conventional techniques. At various times during the period 1996–2004, HLC provided additional information to NRC to refine their computer-based risk analysis, to demonstrate the facility meets the 25 mrem/yr unrestricted release limit of the LTR. The final dose analysis remains incomplete; however, the licensee is expected to provide the additional information in the Fall of 2004.

There is no financial assurance for the facility. HLC remains in operation and is currently revising the estimate of the cost of decommissioning.

## 3.0 MAJOR TECHNICAL OR REGULATORY ISSUES

Current unresolved issues relate to parameters considered in the dose assessment, whether residual contamination in former production areas of the facility should be considered in the building a source term, and whether further investigation is needed in the contaminated floor drain pipe.

## 4.0 ESTIMATED DATE FOR CLOSURE    TBD

# JEFFERSON PROVING GROUND

## 1.0 SITE IDENTIFICATION

Location:           Madison, IN
License No.:        SUB-1435
Docket No.:         04008838
License Status:     Active
Project Manager:    Tom McLaughlin

## 2.0 SITE STATUS SUMMARY

Contamination on site consists of DU in the soil. However, there is a concern for future groundwater contamination. The site has been closed for the testing of all ordnance including DU rounds since 1995. The monitoring of DU in soil, groundwater, surface water, and sediment continues on a bi-annual basis. The U.S. Army submitted a revised DP in June 2002. NRC approved the DP on October 1, 2002.

The licensee currently is proposing to defer decommissioning activities at the JPG site through the establishment of a possession-only license that could be indefinite. Under the proposal, decommissioning will be deferred until the Army can safely collect data needed to validate their off-site transport models. The possession-only license will be issued for a 5-year renewable period, and the status of unexploded ordinance remediation technology will be evaluated at the license renewal to determine if it is appropriate to begin site decommissioning. The Army submitted a license amendment request in September 2003.

There are no immediate radiological hazards at the site. Unexploded ordnance at the site represents a significant non-radiological hazard. The staff has not identified any major off-site environmental issues that will not be addressed during decommissioning of the facility.

The staff does not have an estimate of the cost of decommissioning.

## 3.0 MAJOR TECHNICAL OR REGULATORY ISSUES

The presence of unexploded ordinance, the associated risk, and cost for cleanup of this material, as well as potential contamination of groundwater, are complicating remediation.

The licensee has signed a memorandum of agreement with the Department of the Interior and the Department of Defense (Air Force) for long-term institutional control of the site.

In January 2000, Save the Valley, a local environmental group, requested a hearing on the DP, citing that the DP does not adequately describe the decommissioning process and does not provide adequate assurance for long-term control. The hearing has been extended to include the proposed amendment.

No financial assurance issues have been identified at this time.

## 4.0 ESTIMATED DATE FOR CLOSURE     Indefinite possession-only license

# KAISER ALUMINUM

## 1.0 SITE IDENTIFICATION

Location:              Tulsa, OK
License No.:           STB-472
Docket No.:            040002377
License Status:        Terminated
Project Manager:       John Buckley

## 2.0 SITE STATUS SUMMARY

NRC added Kaiser to the SDMP on August 19, 1994. During site characterization Kaiser identified thorium concentrations above the unrestricted-release limits on Kaiser property and in soil located adjacent to the Kaiser property. Kaiser is remediating the site in two phases. In Phase 1, Kaiser remediated the land adjacent to the Kaiser property. Remediation of the Kaiser property will be performed during Phase 2. Kaiser is requesting unrestricted release of the site.

Phase 1 remediation is complete. Kaiser submitted its FSSR to NRC on August 16, 2001. The staff approved the FSSR March 7, 2002. Kaiser submitted the DP for the Kaiser property (Phase 2) on May 25, 2001. The Phase 2 DP was approved on June 10, 2003.

The NRC staff estimates the cost of decommissioning to be approximately $20 million.

## 3.0 MAJOR TECHNICAL OR REGULATORY ISSUES

On February 12, 2002, Kaiser filed for Bankruptcy (Chapter 11 reorganization). Kaiser has informed NRC that the bankruptcy will not affect ongoing remediation activities at the site.

To date there is minimal public interest in the decommissioning activities at the site. The staff has not identified any major off-site environmental issues that will not be addressed during remediation of the facility.

## 4.0 ESTIMATED DATES FOR CLOSURE     Phase 1 closure – 3/02
                                                                         Phase 2 closure – 5/07

# KERR McGEE – CIMARRON

## 1.0 SITE IDENTIFICATION

Location:           Crescent, OK
License No.:        SNM-928
Docket No.:         07000925
License Status:     Active
Project Manager:    Ken Kalman

## 2.0 SITE STATUS SUMMARY

Contamination at the site consists of uranium contamination in groundwater at Burial Area 1, and Technetium (Tc)-99 in the groundwater in the vicinity of Waste Pond 1 and 2. Concentrations of Tc-99 within applicable release criteria have also been found in Burial Area 1.

The licensee submitted a DP in April 1995 and a DP groundwater evaluation report in July 1998. In coordination with the ODEQ, NRC approved Cimarron's DP in August 1999. Cimarron's license will not be terminated until Cimarron demonstrates that the total uranium concentrations in all wells have been below the groundwater release criteria for eight consecutive quarterly samples.

Most of the site has been released for unrestricted use. Although Subarea G soils meet the release criteria as demonstrated by the NRC staff's confirmatory survey, Subarea G will not be released until there is satisfactory resolution of issues pertaining to the occurrence of Tc-99 in Subarea G.

The site is also licensed for onsite disposal of up to 500,000 cubic feet of Option 2 [of the 1981 Branch Technical Position (BTP)] contaminated soil in Subarea N. NRC staff reviewed Cimarron's Subarea N Report (submitted in January 2002) and performed its independent confirmatory survey in June 2002. Due to a recent occurrence of groundwater exceeding the 180 pCi/l release limit in a nearby portion of Subarea K, NRC is delaying release of Subarea N until the groundwater issue is resolved. Cimarron will not submit its Subarea F FSSR until it has resolved all groundwater issues in that subarea. There are no immediate radiological hazards at the site.

The licensee estimates the cost of decommissioning to be approximately $3.6 million.

## 3.0 MAJOR TECHNICAL OR REGULATORY ISSUES

Groundwater samples have shown concentrations of uranium, Tc-99, fluorides, and nitrates. Tc-99 concentrations appear to be diminishing over time. NRC staff is currently in a dialogue with Cimarron regarding uranium-contaminated groundwater plume emanating from the vicinity of Burial Area 1. Cimarron is considering alternatives for groundwater remediation. ODEQ will retain regulatory control over the non-radiological groundwater components. There is minimal public interest in the decommissioning activities at this site. No financial assurance issues have been identified at this time.

## 4.0 ESTIMATED DATE FOR CLOSURE   5/07

# KERR McGEE – CUSHING REFINERY SITE

## 1.0 SITE IDENTIFICATION

Location:            Cushing, OK
License No.:         SNM-1999
Docket No.:          070-03073
Licensing Status:    Active
Project Manager:     Derek Widmayer

## 2.0 SITE STATUS SUMMARY

Contamination at the site consists of uranium and thorium in the soil and groundwater.

The licensee submitted a DP for the site, in April 1994, that included a request for onsite disposal. The licensee revised the DP on August 17, 1998. The licensee is requesting unrestricted release of the site. In place of onsite disposal, the licensee proposed to ship the waste exceeding the SDMP Action Plan Criteria to Envirocare, for disposal. The staff completed its review of this revised DP (license amendment 10, dated August 23, 1999). The licensee has completed shipping all of its radioactive contaminated waste to Envirocare. The licensee has released portions of the site for unrestricted use (license amendments 13 and 16).

The licensee estimates the remaining cost of decommissioning to be approximately $9.8 million.

## 3.0 MAJOR TECHNICAL OR REGULATORY ISSUES

During a meeting on January 15, 2002, the licensee informed the staff that there is groundwater contaminated with uranium leaving the licensed site. The licensee has developed an ACL for groundwater and has submitted a license amendment to incorporate the ACL for uranium into the license. Staff is completing the licensing actions to incorporate the ACL into the license.

There is moderate public interest in site remediation activities. No financial assurance issues have been identified at this time.

## 4.0 ESTIMATED DATE FOR CLOSURE      12/05

# KERR McGEE TECHNICAL CENTER

## 1.0 SITE IDENTIFICATION

Location:            Oklahoma City, OK
License No:          SUB-986
Docket No:           040-08006
License Status:      Active
Project Manager:     Rachel Browder, R IV

## 2.0 SITE STATUS SUMMARY

The Kerr-McGee Technical Center (KMTC) is a research facility with approximately 50 research laboratories on site. Radioactive contamination of the site mainly stemmed from outdoor test pits where probes for uranium exploration were tested and indoor laboratories where samples of ores were examined. At the KMTC, the licensee had operated a series of calibration test pits containing uranium material, primarily ores and ore concentrates, that had been blended with natural sands to produce dilute known concentrations of uranium and its progeny. The materials which could be present in any form were typically ores containing uranium and thorium, yellowcake ($U_3O_8$), intermediate solid and liquid process streams from a uranium mill, conversion facility and a rare-earths facility, and $UF_6$ in gaseous or liquid form were typically provided in 2 kg cylinders. All of these materials came from licensed fuel cycle facilities. Uranium exploration geological core samples were also tested at this facility. Kerr-McGee Chemical, LLC operations took over the site and used all of its available laboratory space for non-licensed activities.

Termination of the license was requested by letter dated January 7, 1999. KMTC submitted a DP based on ICRP-72 dose factors on April 5, 2001. The DP was approved on June 5, 2003.

The licensee has completed all decommissioning activities at the facility, and subsequently submitted the FSSR for the Outdoor Survey Units on September 15, 2003. The Outdoor Survey Units FSSR was approved on February 9, 2004. On April 15, 2004, the licensee submitted the FSSR for the Indoor Survey Units. The respective report contained pathway analysis modeling for radiological dose assessment from residual radioactive material in embedded and buried piping. The licensee's dose analysis for embedded and buried piping was submitted to NRC and is currently under review.

NRC conducted a final confirmatory survey inspection on May 24–27, 2004, in which 81 surface swipe samples were collected and subsequently sent to the Oak Ridge Institute for Science and Education (ORISE) for analysis. The issuance of the inspection report is pending receipt of the analysis results.

The decommissioning cost estimate provided in the DP was approximately $750,000.

## 3.0 MAJOR TECHNICAL OR REGULATORY ISSUES

None

## 4.0 ESTIMATED DATE FOR CLOSURE      12/04

# KIRTLAND AIR FORCE BASE

## 1.0 SITE IDENTIFICATION

Location:            Albuquerque, NM
License No:          42-23539-01AF
Docket No:           030-28641
License Status:      Indefinitely
Project Manager:     Rachel Browder, R IV

## 2.0 SITE STATUS SUMMARY

Eight radiation training sites were established in November 1961 at Kirtland Air Force Base. The U.S. Government owns the sites. Four of the eight sites, TS5 through TS8, were scheduled for decommissioning after activities at these sites were terminated in 1990. The four inactive training sites were used to train U.S. Department of Defense, Department of Energy, Federal Emergency Management Agency, and other Federal and State personnel to detect dispersed contamination resulting from simulated nuclear weapons accidents.

Thorium oxide sludge was applied and tilled into site soils to simulate dispersed radiological contamination. The thorium oxide sludge served as a low hazard analog for plutonium. A total inventory of 1,710 kilograms of thorium sludge or approximately 602 kg of Th-232 was applied at the sites. Total land area impacted was approximately 44 acres. TS8 was also used as a storage site and had two storage bunkers (Buildings 28005 and 28010) located within its boundary.

A DP was submitted July 14, 2000, and revised November 19, 2002. NRC staff approved the DP on January 6, 2003. NRC Materials License 42-23539-01AF was amended on January 6, 2003, to incorporate the revised DP into License Condition 20.P.

The Air Force permittee has completed all decommissioning activities and is in the process of completing FSS. Portions of TS8, including Building 28005, will remain on the permit and will not be released as originally submitted in the DP. The exact dimension of the new footprint for TS8 has not been determined by the licensee.

The licensee estimates the cost of decommissioning to be approximately $12.8 million.

## 3.0 MAJOR TECHNICAL OR REGULATORY ISSUES

There are no major technical issues. During September 2003, Kirtland Air Force Base was cited for its failure to adhere to standard operating procedures. By letter dated October 21, 2003, the Air Force formally contested the violation. NRC acknowledged the letter on November 26, 2003 and requested additional information from the Air Force. By letter dated May 10, 2004, NRC concluded that the violation was valid and denied the licensee's request that the violation be retracted.

## 4.0 ESTIMATED DATE FOR CLOSURE     4/05

# KISKI VALLEY WATER POLLUTION CONTROL AUTHORITY (KVWPCA)

## 1.0 SITE IDENTIFICATION

Location:          Vandergrift, PA
License No.:       No license
Docket No.:        None
License Status:    Unlicensed
Project Manager:   Ken Kalman

## 2.0 SITE STATUS SUMMARY

Contamination consists of uranium-contaminated sludge ash, with an average concentration of ~147 pCi/g and ~4 percent enrichment distributed in an onsite lagoon. The contamination resulted from the incineration and subsequent re-concentration of effluents released (within regulatory limits) from the nearby Babcox & Wilcox facilities. KVWPCA and its contractors have characterized the contamination in the lagoon with extensive sampling. NRC transmitted site-specific remediation guidance to KVWPCA in November 1999. NRC staff performed a dose assessment for the KVWPCA site and determined that the site meets the LTR criteria for unrestricted release. The staff determined that there is no need for further action and documented its proposal in a draft EA that was published for comment in September 2004. The staff will address any comments received and anticipates issuing a final EA in November 2004.

The NRC staff estimates the cost of decommissioning to be zero.

## 3.0 MAJOR TECHNICAL OR REGULATORY ISSUES

PADEP has taken the position that, under Pennsylvania's Solid Waste Management Act and the ash should be removed from the lagoon. PADEP informed KVWPCA in an April 3, 2003, letter, that disposal of the waste in an appropriately licensed or permitted facility is in the best interest of all parties, that disposal of the waste in a Pennsylvania municipal waste landfill would be prohibited, and that PADEP believes that permanent placement of the ash in the lagoon "would constitute unlawful shallow land burial of low level radioactive waste." NRC staff performed a dose assessment for the KVWPCA site and determined that the site meets the LTR criteria for unrestricted release. Any further actions at the site will be under PADEP's authority.

There is political and public interest about remediation of the KVWPCA site.

## 4.0 ESTIMATED DATE FOR CLOSURE      11/04

# MALLINCKRODT CHEMICAL, INC. (MALLINCKRODT)

## 1.0 SITE IDENTIFICATION

Location:          St. Louis, MO
License No.:       STB-401
Docket No.:        40-6563
License Status:    Active
Project Manager:   John Buckley

## 2.0 SITE STATUS SUMMARY

Contaminants at the Mallinckrodt Chemical, Inc., (Mallinckrodt) site are: U-238; U-235; U-234 and progeny; Th-230; Ra-226; Th-232; Th-228 and progeny; Ra-228; and K-40. Groundwater contamination is not present.

Decommissioning at the Mallinckrodt site will take place in two phases. Phase 1 will decommission the buildings and equipment to the extent that whatever remains on site will be released for unrestricted use. Phase 2 will complete the decommissioning of the building slabs and foundations, paved surfaces, and all subsurface materials to the extent that they can be released for unrestricted use.

Mallinckrodt submitted the Phase 1 DP on November 20, 1997. After several RAIs and several revisions to the DP, NRC approved the Phase 1 DP on May 3, 2002. Remediation at the site began in July 2002. Mallinckrodt submitted its Phase 2 DP on May 15, 2003. Mallinckrodt is requesting to remediate the site to meet the unrestricted release criteria of 10 CFR Part 20, Subpart E.

The estimated cost of decommissioning is approximately $27 million.

## 3.0 MAJOR TECHNICAL OR REGULATORY ISSUES

The Mallinckrodt site has been in operation since 1867 and has produced a wide range of products. In addition to the extraction of columbium and tantalum carried out under NRC license STB-401, various uranium compounds were extracted under contract to the Manhattan Engineering District and the Atomic Energy Commission (MED-AEC). Remediation of MED–AEC radiological constituents is currently being performed under the DOE's FUSRAP by USACE. USACE and Mallinckrodt have yet to agree on who has remediation responsibility for several areas within the facility.

No financial assurance issues have been identified at this time. The staff has not identified any major off-site environmental issues that will not be addressed during decommissioning of the facility. Public interest in the decommissioning activities at the site is moderate.

## 4.0 ESTIMATED DATES FOR CLOSURE    Phase 1 – 1/06, License Termination – 7/08

# MICHIGAN DEPARTMENT OF NATURAL RESOURCES

## 1.0 SITE IDENTIFICATION

Location:            Kawkawlin, MI
License No.:         SUC-1581
Docket No.:          04009015
License Status:      Active
Project Manager:     David Nelson

## 2.0 SITE STATUS SUMMARY

The site covers about 3 acres and is contaminated with thorium. The contamination came from magnesium-thorium alloy production at a defunct former licensee. The contaminated soil is covered with a 1.5 m (5 ft) thick clay cap and encapsulated with 0.9 m (3 ft) thick bentonite slurry walls. Ground water contamination is not an issue at this site.

Michigan Department of Natural Resources submitted the DP on March 3, 2003, with addendums on April 22, 2003. Michigan Department of Natural Resources is requesting unrestricted release of the site. The DP acceptance review indicated insufficient information for a detailed technical review and the DP was rejected in August 2003. On January 30, 2004, Revision 1 of the DP was submitted and the staff anticipates approval in December 2004.

There are no immediate radiological hazards at the site. The staff has not identified any major off-site environmental issues that will not be addressed during decommissioning of the facility.

No financial assurance issues have been identified at this time. The Michigan Department of Natural Resources sought a total of $12.5 million from the Michigan Legislature to complete decommissioning of the site.

## 3.0 MAJOR TECHNICAL OR REGULATORY ISSUES

In 1984, the neighboring licensee undertook encapsulation measures at the site to isolate and prevent the migration of the non-radiological hazardous wastes. Encapsulation measures included the installation of a 1.5m-thick (5 ft) clay cap and 0.9m-thick (3 ft) bentonite slurry walls. As a result, this site involves buried waste that is likely mixed with hazardous chemical wastes. Remediation of the site will require coordination with Michigan Department of Environmental Quality (MDEQ), which regulates hazardous chemicals. The licensee concluded that the mixture of non-radiological hazardous and radioactive waste would make the wastes unacceptable at a chemical or radioactive waste disposal site (other than an authorized mixed-waste disposal facility).

Currently, the State of Michigan does not want the clay cap over the wastes to be removed, because of the non-radiological hazards of the site. However, it is uncertain whether the site can be sufficiently characterized and decommissioned without removal of parts of the cap.

There is minimal, if any, public interest, to date. Public interest is expected to continue to be minimal if the clay cap is not removed and waste removal is kept to a minimum.

## 4.0 ESTIMATED DATE FOR CLOSURE       10/06

# MOLYCORP, INC. – WASHINGTON

## 1.0 SITE IDENTIFICATION

| | |
|---|---|
| Location: | Washington, PA |
| License No.: | SMB-1393 |
| Docket No.: | 040-08778 |
| License Status: | Timely renewal |
| Project Manager: | Tom McLaughlin |

## 2.0 SITE STATUS SUMMARY

Molycorp produced a ferro-niobium alloy from an ore that contained natural thorium with some uranium. The operation resulted in the production of thorium-bearing slag that was used as fill over portions of the site.

Molycorp submitted its original DP in July 1995. After consultation with NRC staff, the licensee stated its intention to submit a revised DP in two parts. Part I of the DP addressed cleanup of the contaminated portion of the site to comply with the SDMP criteria. Part II would address disposal of material from York and Washington in an impoundment on the Washington site and would comply with the LTR. Part 1 of the revised DP was submitted on June 30, 1999. The staff approved the Part I DP on August 8, 2000.

In January 2001, Molycorp withdrew its amendment request for approval of the Part II DP (onsite disposal cell). While Molycorp will continue to decommission the Washington facility under its previously approved Part I DP, it will now dispose of the material off site and will ultimately seek a unrestricted release of the site. On February 26, 2001, Molycorp informed NRC that it finished removal of all its stored above ground waste and shipped the material to the Envirocare facility in Clive, Utah.

Molycorp now has torn down all of its buildings and has sent non-rad contaminated materials off site and rad materials to Waste Control Specialists (WCS). All buildings and foundations have been removed from the site. The licensee has developed and has conducted a new site characterization to determine the amount and extent of contamination and a path forward for decommissioning the surface and subsurface soils.

The licensee estimates the cost of decommissioning to be approximately $30.3 million.

## 3.0 MAJOR TECHNICAL OR REGULATORY ISSUES

Public concern in the Canton Township, City of Washington area, is moderate. Congressional interest also mirrors that found in the local communities.

## 4.0 ESTIMATED DATE FOR CLOSURE    10/06

# NWI BRECKENRIDGE

## 1.0 SITE IDENTIFICATION

Location:            Breckenridge, MI
License No.:         SMB-0833
Docket No.:          040-06264
License Status:      Terminated
Project Manager:     Peter J. Lee, R III

## 2.0 SITE STATUS SUMMARY

Between 1967 and 1970, Michigan Chemical Corporation (MCC) managed the site and used it for the disposal of process wastes from a yttrium recovery operation. These disposal activities were authorized under AEC, License Number, SMB-0833, and were performed in accordance with 10 CFR 20.304, "Disposal by Burial in the Soil." The buried waste material is a solid waste byproduct, known as filtercake, which originated from a rare-earth metal (yttrium) extraction process. Disposal records reported that the filtercake was typically a dense, clay-like material that contained elevated levels of naturally occurring uranium and thorium. After site operations ceased, AEC License Number, SMB- 0833, was terminated.

In addition to the buried wastes, thorium and uranium contaminated surface and subsurface soil has been identified at several locations in open land areas on the site. Several radiological evaluations have been performed in recent years. The most recent of these evaluations took place in November 2001, and led to completion of a characterization report which was submitted to NRC in March 2002. The characterization report provided an estimate for the source term in the area, volume, and the average Th-232 and U-238 concentrations in the buried waste material. The average concentrations were determined to be about 240 pCi/g of Th-232 and 150 pCi/g of U-238.

Environ International Corporation has submitted a "Remedial Work Plan, Waste Excavation and Site Restoration, Breckenridge Disposal Site, St. Louis, Michigan," in March 2004. Region III staff approved the plan in August 2004, and expect remediation to begin in September 2004. Remediation activities are expected to be completed in October 2004.

The licensee estimates the cost of decommissioning to be approximately $750,000.

## 3.0 MAJOR TECHNICAL OR REGULATORY ISSUES

This is an unlicensed site. The Fruit of the Loom Company is the responsible party and is involved in an on-going bankruptcy. The bankruptcy has released sufficient funds for the characterization of the site and disposal of the waste.

## 4.0 ESTIMATED DATE FOR CLOSURE     12/04

# PATHFINDER

## 1.0 SITE IDENTIFICATION

Location:          Sioux Falls, SD
License No.:       22-08799-02
Docket No.:        030-05004
License Status:    Active
Project Manager:   Chad Glenn

## 2.0 SITE STATUS SUMMARY

The Pathfinder Atomic Plant (Pathfinder) operated from August 1966 to September 1967. The nuclear fuel was shipped offsite in 1970 and the plant was placed in SAFSTOR in 1971. In September 1972, Pathfinder's Part 50 operating license was surrendered and the current Part 30 byproduct license was issued. The reactor and fuel storage facilities were decommissioned in 1991 under Reg. Guide 1.86. In November 1992, NRC amended the license to authorize the unrestricted release of the reactor building, fuel storage building, and waste storage building; to demolish the reactor building; and to authorize the possession of fixed activation products at the Pathfinder site.

During its brief operating period a relatively small amount of radioactive contamination was found in the steam turbine and auxiliaries. These systems are collectively referred to as the Balance of Plant (BOP) systems to distinguish it from primary power plant systems such as the reactor and its auxiliaries. The BOP was later decontaminated and disconnected from the reactor plant steam source. The BOP was then integrated into a fossil-fueled peaking plant with gas/oil package boilers suppling steam to operate the existing turbine. The Pathfinder plant that utilized the original nuclear plant's BOP continued to operate on peaking duty until July 13, 2000, when the cooling tower collapsed in a storm. Due to economic reasons, the decision was made to cease operations of the peaking plant. In February 2003, Xcel Energy, the licensee, notified NRC that it had permanently ceased operating activities at Pathfinder. In February 2004, Xcel Energy submitted a DP and license amendment request to authorize decommissioning activities at Pathfinder.

The residual radioactivity is located within plant piping and equipment that comprise the Pathfinder peaking plant. The removal of the radioactive byproduct material within the steam, feedwater, and condensate portions of the BOP is the subject of the Pathfinder decommissioning. According to the license the character of the material is Co-60 (40 mCi) and Zn-65 (1 mCi). This material is in the form of fixed activation products in the BOP.

The estimated cost of decommissioning is approximately $2.8 million.

## 3.0 MAJOR TECHNICAL ISSUES

None

## 4.0 ESTIMATED DATE FOR CLOSURE      4/06

# QUEHANNA (FORMERLY PERMAGRAIN PRODUCTS, INC.)

## 1.0  SITE IDENTIFICATION

| | |
|---|---|
| Location: | Karthaus, PA |
| License No.: | 37-17860-02 |
| Docket No.: | 030-29288 |
| License Status: | Active |
| Project Manager: | John Wray, R I |

## 2.0  SITE STATUS SUMMARY

The Commonwealth of Pennsylvania owns the site and had leased it to Permagrain Products, Inc., (PPI) for the operation of a Co-60 irradiator. After PPI declared bankruptcy in 2002, the license was transferred to the Commonwealth of Pennsylvania in December 2002.

Sr-90 is the main contaminant of concern at the facility, and was used in the manufacture of thermoelectric generators.  Sr-90 contamination is found in buildings as well as in surface and subsurface soil.  Contaminated groundwater is not present at the site. The decommissioning, which started in July 1998 is being performed by Scientech.  Areas which do not meet NRC criteria for unrestricted use were identified as the six hot cells, their respective isolation rooms, two ventilation systems, an overhead crane system, a number of ancillary rooms, and the wastewater treatment building.  Decontamination and demolition of the cell structures was completed in 2004.  Decontamination of the service area floor is complete.  FSS was initiated and is expected to be completed in 2004.  The licensee's FSS plan was submitted in July 2004 and is under review by NRC and ORISE.  ORISE conducted a site familiarization visit in August 2004 in preparation for a confirmatory survey scheduled for October 2004.

The total cost of decommissioning to date has been approximately $25 million.  The licensee expects that an additional $2 million–$3 million will be needed to complete decommissioning activities.

## 3.0  MAJOR TECHNICAL OR REGULATORY ISSUES

In June 2002, the U.S. Department of Justice rejected the Commonwealth's claim that the Federal Government should provide the funding to remediate the site because of a past contract between Martin Marietta and the AEC.  The Commonwealth had informed NRC that the portion of the site containing legacy contamination will be placed into a secure, monitored status until this funding issue is resolved.  In April 2003, the Commonwealth received $7 million from the Federal Government to continue clean up of the facility.  The Commonwealth submitted a license renewal application on March 10, 2003, including a revised DP.  NRC completed its review of the application and renewed the license on September 29, 2003.

Public interest in the decommissioning activities at the site is low.

## 4.0  ESTIMATED DATE FOR CLOSURE     12/04, dependent upon funding

# ROYERSFORD WASTEWATER TREATMENT FACILITY

## 1.0     SITE IDENTIFICATION

Location:              Royersford, Pennsylvania
License No.            non-licensee
Docket No.:            NA
License Status:        NA
Project Manager:       Betsy Ullrich, R I

## 2.0 SITE STATUS SUMMARY

The Royersford Wastewater Treatment Facility (RWTF) receives waste water that contains radionuclides from wastewater generated by a nuclear laundry, UniTech Services Group (UniTech), formerly known as Interstate Nuclear Services (INS). These discharges began in the late 1980's. Elevated levels of radioactivity and radiation have been detected at the RWTF since 1986, in the secondary digestor sludge and the resulting solid products from the dewatering of the secondary digestor sludge. The main contaminants are Co-60 and Cs-137.

Unitech has been in compliance with NRC regulations for disposal to the sanitary sewerage system with typical concentrations of radionuclides in the UniTech wastewater of less than 10% of the regulatory limits for disposal to the sanitary sewer. The total amount of radionuclides released each year, other than tritium and carbon-14, ranges from 66 mCi to 492 mCi. In 2003, UniTech completed installing a pipe from their facility to the Schuylkill River, and obtained a National Pollutants Discharge Elimination System permit for discharge. Royersford has finished cleaning out the lines from UniTech to the RWTF, and cleaned the settling tanks, primary digestor, secondary digestor, etc. The only remaining radioactivity is in the reedbeds, and they are looking into options for the disposal of the reedbed sludge.

To date, no estimate for the cost of decommissioning has been developed.

## 3.0 MAJOR TECHNICAL OR REGULATORY ISSUES

The RWTF secondary digester sludge is a liquid containing 3%–6% solids. Samples of secondary digestor sludge have Co-60 concentrations typically in the range of 9,000–60,000 picocuries per liter (pCi/l), although individual samples have contained as much as 115,000 pCi/l. Cs-137 concentrations are in the range of 1,500–5,000 pCi/l. The disposition of the secondary digestor sludge by mechanical dewatering is performed once or twice each year. The resulting filtercake contains about 20% solids and has been disposed of at a municipal waste landfill. Filtercake samples contain in the range of 22–950 pCi/g for Co-60 and 8–112 pCi/g for Cs-137. Radiation levels measured typically are 80–100 microR/hr near contact with the filtercake.

In 1990, the RWTF began using an onsite reedbed for biological dewatering of secondary digestor sludge. Resulting reedbed sludge is located on site in a 6-foot-high walled reedbed, with the height of the sludge rising as additional material is added. Reedbed sludge is a marsh-like material, containing up to 40% solids. Reedbed sludge samples contained from 77 to 950 pCi/g of Co-60 and 20 to 90 pCi/g of Cs-137. Radiation levels near the surface of the onsite reedbed have increased over time to the range of 800–1000 microR/hr. The reedbed reached its capacity in 2003, and the dried sludge needs to be removed and disposed.

Two main issues are anticipated for closure of the reedbeds: (a) potential radiation doses to workers involved in removal of the reedbed sludge and (a) disposal of the reedbed sludge. Potential dose can be estimated using the licensee's pathway analysis assumption that removal of the sludge would require 10 working days and the average dose rate in the reedbeds. During the year 2000, the average dose rate was 0.345 millirem per hour in the reedbeds; therefore, a person working in the reedbeds for 80 hours could receive 28 millirem from external sources during sludge removal. (The inhalation pathway is a factor of approximately 10,000 smaller, based on the INS pathway analysis.)

Disposal of the reedbed sludge may be more complicated than past disposals of filtercake from mechanical dewatering of sludge to the municipal landfill. In 2000, the Commonwealth of Pennsylvania passed legislation that requires all landfills to have radiation monitors to survey incoming material, to ensure that no radioactive materials are disposed of in the State. The radiation levels from the reedbed sludge will likely be detected by such monitors. The radiation levels from filtercake may also be detected by radiation monitors, but no such disposals have been made since the legislation was passed. Disposal of dried sludge may not meet the requirements of LLW sites, if the sludge contains substances that are considered hazardous materials (this is likely). It is not known at this time if the Commonwealth of Pennsylvania will continue to allow sludge from the RWTF to be transferred to a municipal landfill. The RWTF plans to perform mechanical dewatering in December 2003.

Public interest in the RWTF contaminated sludge is sporadic and is usually associated with issues at the landfill receiving the sludge.

4.0 ESTIMATED DATE FOR CLOSURE     TBD

# SAFETY LIGHT CORPORATION (SLC)

## 1.0 SITE IDENTIFICATION

Location:        Bloomsburg, PA
License No.:     37-00030-02
Docket No.:      030-05980
License Status:  Active
Project Manager: Marie Miller, R I

## 2.0 SITE STATUS SUMMARY

Safety Light Corporation (SLC) is licensed to perform site characterization and decommissioning activities.  Contamination at the site is from the manufacturing operations of self-luminous watch and instrument dials and other items involving Ra-226, Cs-137, Sr-90, and Am-241.  Radioactive waste was disposed on site in three primary locations: silos, lagoons, and a waste dump.  Primary soil contaminates include Ra-226 and Cs-137 with small amounts of Am-241.  The onsite ground water is also contaminated with H-3, Sr-90, and Cs-137.

In October and December 2000, SLC submitted a DP to NRC which called for a "task by task" approach to decommissioning because of limited funding availability.  The DP presents decommissioning activities which will make the site suitable for unrestricted release.  This approach was approved by NRC in December 2001, and on August 15, 2002, NRC amended the SLC license to approve the work plan for processing and sorting waste that was removed from two underground silos in the fall of 1999.

NRC staff continues to coordinate activities with EPA and PADEP regarding remediation of the SLC site.  An EPA Administrative Order of Consent with SLC for the sorting, characterization, and re-packaging of the drums of mixed waste and radioactive waste that were removed from the onsite silos, became effective on February 3, 2003.  A separate EPA Order is expected to be issued in September 2004 for disposal of the waste.  Disposal costs are expected to exceed the licensee's decommissioning funds, so EPA is expected to propose a unilateral Order and use EPA emergency removal funds to complete disposal of the underground silo waste.

The licensee estimates the cost of decommissioning to be approximately $29 million.  An NRC analysis of the licensee's Decommissioning Cost Estimate concluded that the decommissioning cost for unrestricted release of the site by the licensee was estimated to be between $94 million–$120 million and to be $50 million–$78 million for restricted release.

## 3.0 MAJOR TECHNICAL OR REGULATORY ISSUES

Lack of financial assurance remains the key issue.  Effective remediation work cannot be performed because of limited funding.  The licensee's decommissioning fund, which was used to remove and characterize the underground silo waste and ship approximately one-third of this waste, has decreased from $1.9 million to $180,000.

The licenses are due to expire December 31, 2004.  The licensee submitted their request for license renewal, which was received on April 29, 2004.  The letters request a five-year renewal and continuation of the NRC exemption to the financial assurance provisions of 10 CFR 30.

The licensee's renewal applications did not provide sufficient detail, so the staff has developed a RAI. Following receipt of a completed application, the staff will consider the licensee's renewal applications and make a recommendation to the Commission as to whether the present licenses should be renewed. The original settlement agreement indicated that the licenses would not be renewed without adequate financial assurance. However, the staff recommended and the Commission approved renewals in 1999 with an exemption from the financial assurance requirements, with license conditions regarding continued contributions to the decommissioning trust fund and disposal of waste.

In November 2003, Region I was informed by SLC management that required monthly payments to the decommissioning trust fund had not been made for part of 2002 and most of 2003. As a result, on December 19, 2003, NRC issued a Demand For Information (DFI) to SLC because of the licensee's failure to make these payments, which are required by license condition and were specifically approved by the Commission in the 1999 renewal decision. In addition, OI opened a case to address potential deliberate violations in this matter. The licensee responded to the DFI on January 16, 2004. OI issued its report to the staff on March 10, 2004. A predecisional enforcement conference was conducted on July 20, 2004 in Region I. The licensee provided a schedule to make the late payments over the next nine months, and NRC enforcement action is being considered.

In December 2001, NRC requested that EPA Region III conduct a preliminary site assessment for the purpose of scoring the site for inclusion on the National Priorities List (NPL) and possible remediation under CERCLA. EPA has completed the scoring package. The Hazardous Ranking System documentation record was completed in January 2003 and provided to NRC on May 5, 2003. EPA received OMB approval to proceed in April 2004. EPA plans to include SLC on the NPL during its next rulemaking (September 2004).

Public interest in the decommissioning activities at the site is limited.

4.0  ESTIMATED DATE FOR CLOSURE     TBD based on license renewal decision in 12/04

# SALMON RIVER

## 1.0 SITE IDENTIFICATION

Location:           Salmon, ID
License No.:        R-00230 and P-040001
Docket No.:         040003400
License Status:     Terminated
Project Manager:    Dominick Orlando

## 2.0 SITE STATUS SUMMARY

Salmon River Uranium Development (SRUD) is a 21 acre privately-owned site surrounded by U.S. Forest Service lands. Located along the Salmon River, in Salmon, Idaho, the site includes an abandoned mine, a large structure previously used for milling and chemical operations, and a holding pond. The site was licensed from 1958 through 1959. Although both uranium and thorium ore were mechanically and chemically processed at the site, it is suspected that operations with source material were very minor and only experimental in nature.

Despite Idaho Department of Environmental Quality (DEQ) environmental evaluations in the early 1990's, NRC was not involved with the site from 1962 until the site was identified as part of the terminated license review project. On May 22, 2001, staff from NRC Region IV visited the former SRUD site and identified thorium contamination in the form of partially processed ore. Laboratory results confirmed that the material onsite was "source material" (i.e., >0.05 wt% thorium).

On July 3, 2001, the property owners were notified regarding the results of the site inspection. NRC staff have reviewed the docket and inspection results and determined that (a) some level of remediation will be required at the site; (b) the site should be remediated to LTR rather than Part 40 Appendix A criteria; and (c) based on the available information at the site, the cost of remediation (based on very limited surveys) could be approximately $70,000.

A site inspection and scoping survey was performed at the site in October 2003, and confirmed that residual contamination is present in two site structures and in soil around the mill site. Staff is working with the site owners, the State, and the U.S. Forest Service to develop a path forward for the site.

## 3.0 MAJOR TECHNICAL OR REGULATORY ISSUES

Financial
The site is non-licensed and therefore has no financial assurance. Because the site is owned by a private citizen (rather than a corporate entity) it is currently not clear whether the owner is financially capable of funding site remediation.

Interagency Coordination
For several reasons this site may require interagency coordination to achieve cleanup (e.g., Idaho DEQ has already overseen a chemical cleanup at the site, the site is surrounded by Federal U.S. Forest Service lands, an EPA emergency action could be required at the site because of the waste stored in removable bags, etc.).

## 4.0 ESTIMATED DATE FOR CLOSURE     TBD

# SCA SERVICES (SCA)

## 1.0  SITE IDENTIFICATION

Location:            Kawkawlin, MI
License No.:         SUC-1565
Docket No.:          04009022
License Status:      Active
Project Manager:     David Nelson

## 2.0  SITE STATUS SUMMARY

A portion of the site is contaminated with thorium from magnesium-thorium alloy production at a defunct former licensee.  The contaminated soil is covered with a clay cap and encapsulated with slurry walls and in two small piles covered with clay.  There are also hazardous wastes present at the site.  Site characterization including the potential for ground water contamination is being evaluated.  The site is being regulated under the State Superfund law.  NRC issued a license amendment on October 10, 2001, extending the submittal date of the DP to September 30, 2003.  A DP dated November 2003 was submitted, and NRC accepted the DP for review in a letter dated March 1, 2004.  The licensee is requesting unrestricted release of the site.

There are no immediate radiological hazards at the site.  The staff has not identified any major off-site environmental issues that will not be addressed during decommissioning of the facility.

The estimated cost for decommissioning the site is $1.9 million.

## 3.0  MAJOR TECHNICAL OR REGULATORY ISSUES

The licensee undertook cap repair measures at the site to isolate and prevent the migration of the non-radiological hazardous wastes.  Remediation of the site will require coordination with MDEQ, which regulates hazardous chemicals.  The mixture of non-radiological hazardous and radioactive waste would make the wastes unacceptable at a chemical or radioactive waste disposal site (other than an authorized mixed-waste disposal facility).  Currently, the State of Michigan does not want the clay cap over the wastes to be removed, because of the non-radiological hazards of the site.

There is minimal, if any, public interest to date.  Public interest is expected to remain minimal if the clay cap is not removed.

No financial assurance issues have been identified to date.

## 4.0  ESTIMATED DATE FOR CLOSURE     7/11

# SHIELDALLOY METALLURGICAL CORPORATION (SMC)

## 1.0 SITE IDENTIFICATION

Location:          Newfield, NJ
License No.:       SMB-1507
Docket No.:        04007102
Licensee Status:   Active
Project Manager:   Ken Kalman

## 2.0 SITE STATUS SUMMARY

Contamination at the Shieldalloy Metallurgical Corporation (SMC) site is in the form of facility-generated slag and baghouse dust. The major contaminants are natural uranium and natural thorium. The site is also on the NPL under CERCLA, because of past operations involving chromium-contaminated onsite groundwater. Remediation of the groundwater is currently taking place.

In August 2001, SMC notified NRC that they had ceased production activities using source material. On August 27, 2001, the licensee provided notification and intent to decommission. The license is in timely renewal, and was amended on November 4, 2002, to authorize only decommissioning activities that were previously permitted. The licensee submitted a revised license renewal application on May 1, 2003.

SMC submitted its DP on August 30, 2002. The DP was rejected and on May 16, 2003. NRC staff is using a phased approach to enable SMC to develop an acceptable DP for a possession only license for long term control of the site. As part of this phased approach, in May 2004, the NRC staff provided guidance to SMC regarding the use of a Possession only long term control license. The New Jersey Department of Environmental Protection has taken issue with the use of a long term control license, as stated in a letter to the Chairman dated May 15, 2004. NRC staff has prepared a response to this letter for the Chairmans signature and anticipates transmitting the letter in August 2004.

The licensee estimates the cost of decommissioning to be approximately $1.8 million dollars.

## 3.0 MAJOR TECHNICAL OR REGULATORY ISSUES

In the past, SMC has found it difficult to sell the slag material. Several attempts to export the material have failed. SMC intended to sell the baghouse dust to a local cement manufacturer, however, no buyer has been found. Regardless of whether the sales occur, SMC has proposed to dispose of these materials on site in an engineered cell.

SMC has less than adequate financial assurance for decommissioning. To date, public interest in the decommissioning activities of this site is minimal.

## 4.0 ESTIMATED DATE FOR CLOSURE     9/10

# STEPAN CHEMICAL COMPANY

## 1.0 SITE IDENTIFICATION

Location:           Maywood, NJ
License No.:        STC-1333
Docket No.:         40-8610
License Status:     Active
Project Manager:    Amir Kouhestani

## 2.0 SITE STATUS SUMMARY

Principal radioactive contaminants at the site are process wastes from the thorium extracted from the monazite sands using a chemical separation process. These alkaline thorium phosphate tailings are stored in three licensed underground storage areas.

The license is in timely renewal. The site is on the EPA NPL and is also designated as a FUSRAP site. The USACE has the lead authority for cleanup of FUSRAP sites under CERCLA of 1980, as amended. The EPA has an oversight role at this site due to a Federal Facility Agreement signed between DOE (USACE's FUSRAP predecessor) and the EPA. The cleanup of the licensed burials is part of a ROD for Soils and Buildings at the FUSRAP Maywood Superfund Site (August 2003). The USACE prepared the ROD in accordance with the National Contingency Plan requirements, with public input, and in consultation with EPA, New Jersey Department of Environmental Protection, and NRC. In July 2001, to avoid dual regulation of site cleanup activities, NRC and USACE entered into a MOU for coordination of decommissioning of FUSRAP sites with NRC-licensed facilities. Site cleanup will be guided by the NRC–USACE FUSRAP MOU. The MOU provides that if certain NRC conditions are met by USACE and the licensee, NRC will put the Stepan license in abeyance while USACE is conducting site remediation. Upon completion of site cleanup, NRC will reinstate the license, and if satisfied with USACE cleanup, NRC will proceed with license termination subject to the effectiveness rules of the NRC hearing process pursuant to 10 CFR Part 2. In January 2004, the licensee submitted a draft proposal for a USACE/licensee request for license suspension. NRC is reviewing the proposal.

The MOU and the ROD commit USACE to clean up the licensed burials to meet at least the NRC standards required under 10 CFR 20.1402; Radiological criteria for unrestricted use, or a more stringent criteria.

A final decommissioning cost estimate is currently being negotiated.

## 3.0 MAJOR TECHNICAL OR REGULATORY ISSUES

The decommissioning schedule is in major part contingent upon a settlement on Potential Responsible Party liabilities under CERCLA, the FUSRAP Appropriations, and USACE's greater FUSRAP Maywood Superfund project schedule and funding priorities. The staff has not identified any major offsite environmental issues that will not be addressed during decommissioning of the facility.

## 4.0 ESTIMATED DATE FOR CLOSURE      9/09

# SUPERIOR STEEL (FORMERLY SUPERBOLT)

## 1.0 SITE IDENTIFICATION

Location:            Carnegie, PA
License No.:         N/A – non licensed facility
Docket No.:          N/A – non licensed facility
License Status:      Terminated
Project Manager:     Robert Prince, R I

## 2.0 SITE STATUS SUMMARY

Superbolt, the current owner of the facility, is not a licensee. The site was owned and operated by the Superior Steel Company, during the period from 1952 to 1957. During this period Superior Steel performed contract work for the AEC. Superior Steel's license expired in 1958.

The Superbolt site consists of five interconnected warehouse buildings with uranium contaminated building surfaces. Uranium contamination is also present in a sub-floor trench located within two of the warehouse buildings and extending approximately 50 feet outside the building structure. The trench has been backfilled and sealed with a layer of concrete. Historical surveys indicated the presence of ground water intrusion in portions of the sub-floor trench. However, no indication of ground water contamination beyond the trench boundary has been detected. Uranium contamination was also detected outside and adjacent to the building.

Currently no DP exists for the site. The licensee does not have a cost estimate for decommissioning. Based on characterization survey data and historical information it is believed that the major cost contributor will be the cost associated with excavation of the material in the sub-floor trench and associated disposal costs that could be several million dollars or higher.

## 3.0 MAJOR TECHNICAL OR REGULATORY ISSUES

Funding of remediation work is the primary concern. Superbolt is a small company with limited financial resources at their disposal. The company reported a net loss in fiscal year ending September 2002 and a net income of less than $150,000 for fiscal year ending September 2003. Superbolt may proceed with limited remediation work, if their financial condition allows, to gain use of some of the building areas to expand production. Additional production capacity is crucial if Superbolt is to improve its financial position. However, company officials have expressed concern that any effort on their part to remediate building areas could be interpreted as accepting responsibility for the contamination and the associated liability. Based on this concern and their financial condition company officials are hesitant to move forward with remediation efforts.

The State of Pennsylvania recently requested that DOE reconsider the site for funding under FUSRAP. This request was denied, and the State is currently pursuing other avenues for funding. Because the sub-floor trench does not pose any radiological concerns to individuals working in the buildings, efforts could be directed towards remediation of building areas for productive use by Superbolt. Long term options could include restricted release of the site.

Superbolt expects to retain ownership of the facilities and has no plans at this time to sell the property.

4.0 ESTIMATED DATE FOR CLOSURE    TBD—pending resolution of funding

# UNC NAVAL PRODUCTS

## 1.0  SITE IDENTIFICATION

Location:           New Haven, CT
License No.:        SNM-368
Docket No.:         070-00371
License Status:     Terminated
Project Manager:    Marie Miller, R I

## 2.0  SITE STATUS SUMMARY

This site had been used by United Nuclear Corporation (UNC) to fabricate nuclear fuel components for the U.S. Government, was decommissioned in 1976, and removed from NRC License No. SNM-368 on April 22, 1976.  As part of the ORNL review of Terminated Licenses, NRC conducted independent measurements in May 1996, and additional measurements on September 1996.  Results indicated residual enriched uranium exceeding the 30 pCi/g in soil or sediments in two buildings and a connected inactive sewer system.  The site is on the edge of an industrial redevelopment area, and the buildings, which are enclosed by a chain link fence, are presently used as a warehouse by the property owner.

The contaminated areas were documented in a July 1996 NRC Inspection Report.  In June 1998, the licensee agreed to characterize and remediate the facility in accordance with Option 1 delineated in the NRC BTP for "Disposal or Onsite Storage of Thorium or Uranium Wastes from Past Operations."  The licensee submitted a characterization plan and DP in August 1999, and conducted sampling activities in 2003.  A final Characterization Plan is expected to be submitted in December 2004.

The former licensee is completing characterization to determine the cost of decommissioning.

## 3.0  MAJOR TECHNICAL OR REGULATORY ISSUES

Because UNC had maintained an open contract with the U.S. government, there was no site remediation until the cost recovery was resolved with DOE.  In addition, site radiological activities have to be contracted through a competitive contract process that has caused some delays.  Further, UNC, as a non-NRC licensee, is not required to comply with the Decommissioning Timeliness Rule.  Also, UNC is not the owner of the site, but has agreed to undertake the remediation.

The State of Connecticut and the City of New Haven have some interest in the site, because it is part of a redevelopment area.  The general public has relatively low interest in the site.  The sewer authority has cooperated with NRC and UNC in collection of sewer samples.

## 4.0  ESTIMATED DATE FOR CLOSURE     TBD

# UNION CARBIDE CORPORATION

## 1.0 SITE IDENTIFICATION

Location:            Lawrenceburg, TN
License Nos.:        SNM-724, SMB-720
Docket Nos.:         070-00784, 040-07044
License Status:      Terminated
Project Manager:     Ken Kalman

## 2.0 SITE STATUS SUMMARY

The contaminant at the Union Carbide site is enriched uranium. Uranium contamination is present in buildings and soil. Ground water contamination is not an issue at this site.

The UCAR Carbon Company, Inc., (UCAR) DP was approved in two phases: Phase 1, decommissioning activities associated with buildings was approved on July 27, 2000; Phase 2, decommissioning activities associated with soil was approved on December 1, 2000. UCAR is using the cleanup criteria found in the 1993 "Guideline for Decommissioning of Facilities" for buildings and structures. UCAR is "grand-fathered," and thus able to use these criteria for buildings. This decommissioning approach will allow release of the site for unrestricted use.

There are no immediate radiological hazards at the site.

The NRC staff estimates the cost of decommissioning to be approximately $600,000.

## 3.0 MAJOR TECHNICAL OR REGULATORY ISSUES

No financial assurance issues have been identified to date. Public interest about decommissioning activities at the site is minimal. The staff has not identified any major off-site environmental issues that will not be addressed during decommissioning of the facility.

## 4.0 ESTIMATED DATE FOR CLOSURE      12/05

# WEST VALLEY

## 1.0 SITE IDENTIFICATION

| | |
|---|---|
| Location: | West Valley, NY |
| License No: | CSF-1 |
| Docket No.: | 0500201, POOM-032 |
| License Status: | In abeyance |
| Project Manager: | Chad Glenn |

## 2.0 SITE STATUS SUMMARY

The West Valley site is located on the Western New York Nuclear Service Center (Center) and comprises 3,300 acres of land established for siting a former reprocessing facility. The New York State Energy Research and Development Authority (NYSERDA) holds title to this land. In its regulatory responsibilities under the Atomic Energy Act, the AEC, and subsequently NRC, licensed (CSF-1) the site from 1966 to 1981. The Center contains a nuclear fuel reprocessing facility that operated from 1966 to 1972, and produced approximately 600,000 gallons of liquid HLW. The Center also contains contaminated structures and two radioactive waste disposal areas: (1) a 15-acre New York State–licensed disposal area (SDA) that operated as a commercial LLW disposal facility from 1963 to 1975, and (2) a 5-acre NRC-licensed disposal area (NDA) that received radioactive wastes from the reprocessing plant and associated facilities from 1966 through 1986. In addition to the reprocessing facility and disposal areas, the Center includes a HLW tank farm, waste lagoons, above-ground radioactive waste storage areas, with soil and groundwater contamination near these facilities.

In 1980, Congress enacted the WVDP Act. Under the Act, DOE assumed exclusive possession of the 200-acre portion of the Center which includes the former reprocessing facility, the NDA, the HLW tanks, waste lagoons, and above-ground waste storage areas. The WVDP Act authorized DOE to: solidify, transport and dispose of HLW that exists at the site; dispose of LLW and transuranic waste produced by the WVDP; and decontaminate and decommission facilities used for the WVDP in accordance with requirements prescribed by NRC. In 1981, NRC put the license in abeyance to allow DOE to carry out the WVDP Act. In 2002, DOE completed the solidification of liquid HLW which was placed into 275 stainless steel canisters. The HLW canisters are expected to be stored onsite until shipped for disposal to the geologic repository.

In 2002, the Commission issued its final policy statement on decommissioning criteria for the WVDP. The policy statement prescribed the LTR as the decommissioning criteria for the WVDP, reflecting the fact that the applicable goal for the entire NRC-licensed site is compliance with the requirements of the LTR.

DOE and NYSERDA are developing a Decommissioning and Long-term Stewardship EIS. NRC is a cooperating agency for this EIS in accordance with its responsibilities under the WVDP Act. NRC intends to use this EIS to fulfill its NEPA responsibilities for applying the LTR to the WVDP and to assist in its determination of whether the preferred alternative meets the LTR. DOE and NYSERDA are working to define the alternatives to be analyzed in the EIS. DOE's current schedule forecasts public release of the draft EIS in October 2006, and release of the final EIS in August 2007.

DOE also plans to submit a DP for the WVDP to NRC. In this DP, DOE intends to evaluate residual radiological contamination across the entire West Valley site. DOE presently intends to submit its DP to NRC in 2005.

3.0 MAJOR TECHNICAL OR REGULATORY ISSUES

- Long-term site stewardship
- Payment of HLW disposal fees
- Waste Incidental to Reprocessing
- Effects of erosion on disposal areas

4.0 ESTIMATED DATE FOR CLOSURE     TBD

# WESTINGHOUSE ELECTRIC COMPANY

## 1.0 SITE IDENTIFICATION

Location:           Blairsville, PA
License Nos.:       SNM-37, SUC-509
Docket Nos.:        070-00026, 040-03558
License Status:     Terminated
Project Manager:    Mark Roberts, R I

## 2.0 SITE STATUS SUMMARY

The plant was founded in 1955 as a R&D and manufacturing facility for Westinghouse. AEC License Nos. SNM-37 and SUC-509 authorized commercial and naval fuel manufacturing and R&D utilizing low-enriched, high-enriched, and DU. These licenses expired or were terminated on July 1, 1961, and December 31, 1964, respectively. The facility is currently involved in zircalloy tubing manufacturing operations and radioactive materials are no longer used.

Under the terminated license review project, Westinghouse personnel performed radiological measurements in interior areas and identified non-removable surface contamination at levels exceeding current NRC guidelines for release for unrestricted use. Uranium contamination was also identified in two sumps and a number of drain lines beneath the concrete floor. Measurements outside the facility indicated soil contamination exceeding the criteria for release for unrestricted use in a construction material dump east of the Main Building and at the site of a former waste processing building and zircalloy burn area.

In December 1994, Westinghouse commenced remediation of contaminated surfaces and removed the sumps and interior drain lines. Remediation activities were scheduled so that manufacturing operations were not impacted. Characterization and removal of contaminated soil were completed in the dump area in the fall of 1997. The area surrounding the former waste processing building and zircalloy burn area was remediated in 2001. Waste from remediated areas was disposed at the Envirocare facility in Utah.

Westinghouse has prepared and submitted documentation on the surveys performed in the interior areas. Surveys for the exterior areas have been performed, but final survey documentation have not been submitted to the NRC staff. Monitoring well data do not indicate any groundwater issues.

Other than report preparation, the licensee does not expect any additional costs for decommissioning.

## 3.0 MAJOR TECHNICAL OR REGULATORY ISSUES

None

## 4.0 ESTIMATED DATE FOR CLOSURE     12/04

# WESTINGHOUSE ELECTRIC COMPANY (HEMATITE FACILITY)

## 1.0  SITE IDENTIFICATION

Location:          Festus Township, Jefferson County, MO
License No.:       SNM-33
Docket No.:        07000036
License Status:    Active
Project Manager:   Amir Kouhestani

## 2.0  SITE STATUS SUMMARY

The property consists of approximately 228 acres.  The operating facility consists of two main plant buildings, an administration and several support buildings, and a parking area.  Plant operations included low-enriched uranium fuel fabrication, processing and treating uranium compounds, including all forms of uranium from depleted to enriched uranium, and thorium fuel. Contamination at the site consists of uranium and thorium in the soil and groundwater.

The Westinghouse Electric Company, LLC (WEC) has provided phased-notification of cessation of operational activities.  On September 11, 2001, WEC provided notification of cessation of all principal activities and submitted an application for license amendment to change the scope of authorized license activities to those associated with decommissioning activities.  WEC has performed, within its permitted license activities, certain equipment decontamination and dismantlement and has shipped equipments and material to its South Carolina facility.  NRC approved an alternate decommissioning schedule to submit a site-wide DP by April 30, 2004.

On September 28, 2000, the Missouri Department of Natural Resources, under the authority of CERCLA through a cooperative agreement with the EPA, issued a Site Inspection Report of the facility.  The Missouri Department of Natural Resources report suggests that radioactive wastes were disposed on site, resulting in onsite soil and groundwater contamination.  On November 26, 2003, the Missouri Department of Natural Resources approved WEC's Remedial Investigation/Feasibility Study Work Plan dated May 9, 2003.

WEC submitted the Phase 1 DP in April 2004, and Decommissioning Funding Plan is anticipated in September 2004.  The staff anticipates approval of the Phase 1 DP by December 2005.

Decommissioning is estimated to cost approximately $40.5 million.

## 3.0 MAJOR TECHNICAL OR REGULATORY ISSUES

Although no financial assurance issues have been identified at this time, WEC plans to submit an  updated Decommissioning funding plan each time a new DP phase is submitted.  The phased DP submittal and coordination of remedial investigation and remedial action under CERCLA versus NRC's LTR decommissioning and site cleanup criteria could potentially be challenging.  The staff has not identified any major offsite environmental issues that will not be addressed during decommissioning of the facility.

There are active local, State, and Congressional interests in how the site will be decommissioned.

4.0  ESTIMATED DATE FOR CLOSURE     3/10

# WESTINGHOUSE ELECTRIC COMPANY, WALTZ MILL

## 1.0 SITE IDENTIFICATION

Location:            Madison, PA
License No.:         SNM-770
Docket No.:          070-00698
License Status:      Active
Project Manager:     Mark Roberts, R I

## 2.0 SITE STATUS SUMMARY

The Waltz Mill site is currently licensed primarily to provide testing, calibration, and maintenance services for contaminated reactor servicing equipment and other reactor components. Radiological contamination in soil and groundwater exist on a portion of the site as a result of the clean-up activities following a 1961 incident at the TR-2 test reactor, waste segregation activities, and nuclear laundry services. Significant contamination was also present in retired facilities (hot cells, hot cell support rooms, and a section of the fuel transfer canal) within one of the site buildings. Contaminants are primarily Sr-90 and Cs-137, with lesser quantities of mixed fission, activation products, and trace levels of transuranic radionuclides. Due to a series of corporate mergers, the licensee for the TR-2 test reactor is Viacom, Inc.

Westinghouse submitted a remediation plan in April 1997. NRC approved the remediation plan in January 2000. The licensee has remediated much of the interior and exterior contaminated areas. Remediation activities focused on the three hot cells and supporting facilities in conjunction with work on decommissioning the test reactor. Contaminated soil removal has been completed in the primary exterior remediation area, although small pockets of contaminated soil have been identified on the site.

Westinghouse and Viacom have a total of $30 million in financial assurance agreements for completion of site decommissioning.

## 3.0 MAJOR TECHNICAL OR REGULATORY ISSUES

The licensee does not intend to request termination of the license and is expanding site operations through consolidation of business from other sites. The licensee went forward with the remediation project, in part, to address the reasons why the facility was originally placed on the SDMP list. The Viacom TR-2 license was intended to be terminated following decommissioning of the test reactor and the building transferred to the Westinghouse SNM-770 license. Westinghouse and Viacom have not reached an agreement on the transfer. This issue and related issues are currently the subject of a commercial dispute that is being resolved under arbitration.

PADEP has interest in the condition of the site, particularly groundwater issues. No financial assurance issues have been identified at this time.

## 4.0 ESTIMATED DATE FOR CLOSURE     8/05

# WHITTAKER CORPORATION

## 1.0 SITE IDENTIFICATION

Location:            Greenville, PA
License No.:         SMA-1018
Docket No.:          040-7455
License Status:      Active
Project Manager:     Craig Gordon, R I

## 2.0 SITE STATUS SUMMARY

Whittaker's license authorizes possession of licensed material for storage only.  Thorium is the most abundant contaminant onsite, along with smaller amounts of uranium and radium contamination in soil.  A license renewal application was received in May 2004.

Decommissioning activities began in July 2004, by Whittaker's contractor with site cleanup of non-radiological wastes and disposal of some stored radioactive wastes in a licensed waste disposal facility.  Selected decommissioning activities are being performed under the current license.  In addition, a license amendment request was received in May 2004 for crushing and blending contaminated soil and slag for offsite disposal.  The proposal is under NRC review, andis expected to be approved in the fall of 2004.  A dose assessment was provided which shows doses from transportation of material will be below the LTR.

A separate request was also submitted to NRC for review of site-specific DCGLs for residual materials left onsite.  The licensee expects to resubmit this request with additional dose assessment analyses.

The licensee will continue with plans for blending waste material to support license termination and unrestricted release.  Remediation work which began July 2004 is expected to be completed in July 2005, as proposed in Whittaker's current schedule.

Whittaker's initial estimates for decommissioning costs was approximately $6.7 million for unrestricted release.

## 3.0 MAJOR TECHNICAL OR REGULATORY ISSUES

Whittaker is actively investigating beneficial reuse of non-source material slag recovered from soil remediation activities.  The potential for beneficial reuse can affect total decommissioning costs and the amount of funds necessary for financial assurance.

Public interest in the decommissioning activities at this site is low.

## 4.0 ESTIMATED DATE FOR CLOSURE     9/05

# Appendix D

## Site Summaries for Title II Sites Undergoing Decommissioning

# AMERICAN NUCLEAR CORPORATION

## 1.0  SITE IDENTIFICATION

| | |
|---|---|
| Location: | Gas Hills, Fremont County, WY |
| License No.: | SUA-667 |
| Docket No.: | 40-4492 |
| License Status: | Reclamation |
| Project Manager: | John H. Lusher |

## 2.0  SITE STATUS SUMMARY

Reclamation oversight of the facility has been transferred to the State of Wyoming, Department of Environmental Quality (WDEQ).  This transfer occurred because ANC had become insolvent in May 1994 and site reclamation was incomplete.  A Confirmatory Order between NRC and the WDEQ describing the requirements for reclamation activities was agreed upon by both parties and was issued in October 1996.

The licensed site encompasses approximately 550 acres of which approximately 80 acres consist of Tailings Pile 2 and 40 acres of Tailings Pile 1.  Tailings Pile 2 reclamation activities were completed and approved by NRC in February 1998 and Tailings Pile 1 activities are on hold.  Additionally, the site has an active groundwater recovery and corrective action program.

Reclamation activities are targeted to restart in 2005.  Since the last inspection in July 2002, no site reclamation activities had been performed.  After approval of the reclamation plan for Tailings Pile 1, activities will include the following:  (1) windblown area clean-up activities, (2) capping with clay, (3) radon testing, and (4) placement of rip-rap rock.

WDEQ's goal is to submit the final reclamation plan for Tailings Pile 1 for NRC review and approval in 2004 and complete the associated reclamation in 2005.

The cost for decommissioning is estimated to be approximately $3.2 million.  WDEQ has approximately $3.2 million in DOE Title10 funds to complete reclamation of Pile 1.

## 3.0  MAJOR TECHNICAL AND REGULATORY ISSUES

None

## 4.0  ESTIMATED DATE FOR CLOSURE     2007

# BEAR CREEK

## 1.0 SITE IDENTIFICATION

Location:           Converse County, WY
License No.:        SUA-1310
Docket No.:         40-8452
License Status:     Reclamation
Project Manager:    Rick Weller

## 2.0 SITE STATUS SUMMARY

The decommissioning and reclamation of the Bear Creek Uranium Mill, including the mill tailings impoundment, was completed in November 1999. The tailings impoundment contains 4.7 million tons of uranium ore tailings and covers an area of approximately 101 acres. The staff performed a final "closeout" inspection of the completed reclamation construction activities in July 2000. The staff completed its review of the Bear Creek Tailings Reclamation Construction Report in July 2001 with the conclusion that reclamation of the Bear Creek tailings impoundment was completed in accordance the requirements of 10 CFR Part 40, Appendix A, and the licensee's approved Tailings Reclamation Plan. At the Bear Creek site, the State of Wyoming owns both the surface estate (where the tailings impoundment is located) and the subsurface estate with the contained mineral rights. The licensee purchased the surface estate from the State in January 2003 and is currently negotiating with the State to acquire the subsurface estate. Following the successful acquisition of the subsurface estate, the licensee can prepare the necessary papers to turn over ownership of the site to the DOE for long-term custody and subsequent termination of its NRC license.

The cost for decommissioning is estimated to be approximately $900,000.

## 3.0 MAJOR TECHNICAL AND REGULATORY ISSUES

There is one major regulatory issue at the Bear Creek site that must be addressed as part of the license termination process. That issue relates to the current State ownership of the subsurface estate at the site. If the State does not divest itself of ownership of the subsurface estate, the State will become the long-term custodian for those interests. However, the State has previously expressed its disinterest in becoming the long-term custodian for any Title II (Uranium Mill Tailings Radiation Control Act [UMTRA] of 1978, as amended) site in Wyoming. Accordingly, licensee and State representatives are currently negotiating an exchange of mineral interests (the subsurface estate) so that the licensee can acquire ownership of the entire Bear Creek site, including the subsurface interests. License termination is pending (a) a successful exchange of mineral estates between the licensee and the State, (b) transfer of the Bear Creek site property by the licensee to the DOE, (c) payment of the long-term surveillance fee by the licensee, and (d) NRC approval of the LTSP currently being prepared by the DOE.

## 4.0 ESTIMATED DATE FOR CLOSURE     2004

# COGEMA MINING, INC.

## 1.0 SITE IDENTIFICATION

Location:          Northeastern (Johnson and Campbell Counties) Wyoming
License No.:       SUA-1341
Docket No.:        40-8502
License Status:    Reclamation
Project Manager:   Elaine Brummett

## 2.0 SITE STATUS SUMMARY

The area of these two in-situ leach sites (7 miles apart) disturbed by well fields or facilities is approximately 687 acres. Groundwater restoration is complete at Irigaray and nearly complete at Christensen Ranch. Surface decommissioning began in 2002 (plan approved December 31, 2001), and there are minor amounts of soil contamination. Equipment and building components that cannot be decontaminated will be shipped with the contaminated soil to the Pathfinder – Shirley Basin tailings impoundment for disposal.

The cost for decommissioning is estimated to be approximately $12.1 million.

## 3.0 MAJOR TECHNICAL AND REGULATORY ISSUES

None

## 4.0 ESTIMATED DATE FOR CLOSURE     2007

# EXXONMOBIL HIGHLANDS

## 1.0 SITE IDENTIFICATION

| | |
|---|---|
| Location: | Converse County, WY |
| License No.: | SUA-1139 |
| Docket No.: | 40-8102 |
| License Status: | Reclamation |
| Project Manager | John H. Lusher |

## 2.0 SITE STATUS SUMMARY

The Highland uranium recovery facility included a conventional surface uranium mine with an associated mill. The site also included ore storage pads, four pit mines (two of which have been backfilled), several waste rock piles, one tailings impoundment, and an environmental laboratory.

Surface mining (beginning in 1970), solution mining (beginning in 1972) and underground mining (beginning in 1977) were used to recover the uranium ore. The first ore was processed at the facility in October 1972. Approximately 11.3 million tons of ore were processed at the Highland mill using an acid leach circuit between 1972 and the end of operations in mid-1984. The resulting tailings consisted of sand and slime fractions, which were discharged to the tailings basin. Solution contained in the basin was occasionally recycled to the mill, and pilot aquifer restoration waters were discharged to the tailings basin.

The uranium mill area, including the ore storage pads and the laboratory have been cleaned up and the tailing and wind blown tailings areas are buried under the radon barrier, eliminating nearly all potential for radiation exposures to workers or members of the general public from these sources. All construction operations for the tailings reclamation were completed in 1991. The NRC staff reviewed the construction completion report and found it acceptable in September 2002.

ExxonMobil needs to determine the extent of the reclamation site boundary and establish and prepare all land transfers to DOE. DOE needs to prepare the draft LTSP for NRC review and comment and then provide the final LTSP.

The cost for decommissioning is estimated to be approximately $1.3 million.

## 3.0 MAJOR TECHNICAL AND REGULATORY ISSUES

None

## 4.0 ESTIMATED DATE FOR CLOSURE    2005

# HOMESTAKE

## 1.0  SITE IDENTIFICATION

Location:              Grants, NM
License No.:           SUA-1471
Docket No.:            40-8903
License Status:        Reclamation
Project Manager:       William von Till

## 2.0  SITE STATUS SUMMARY

The facility is a conventional uranium mill site under reclamation.  Uranium processing started in the late 1950's and continued until 1990.  Tailings generated from the milling operation were placed on two piles, a large pile and a small pile.  The facility has a tailings area of 170 acres with a weight of 22 million tons.  Currently there are several evaporation ponds and an ion exchange treatment building for groundwater remediation, and several administrative and maintenance buildings.  Seepage from the tailings piles was noted in 1975.

The current effort is a major groundwater corrective action plan which is also under the oversight of the EPA through Superfund.  A MOU has been executed between NRC and EPA for this site regarding groundwater remediation.  Staff is currently reviewing a license amendment to revise groundwater compliance standards.  In accordance with the MOU, staff is consulting with EPA and the State of New Mexico to review this action.

The cost for decommissioning is estimated to be approximately $35.3 million.

## 3.0  MAJOR TECHNICAL AND REGULATORY ISSUES

None

## 4.0  ESTIMATED DATE FOR CLOSURE     2015

# PATHFINDER – LUCKY MC

## 1.0  SITE IDENTIFICATION

Location:            Gas Hills Mining District, WY
License No.:         SUA-672
Docket No.:          40-2259
License Status:      Reclamation
Project Manager      Elaine Brummett

## 2.0  SITE STATUS SUMMARY

The Lucky MC site is located in west central Wyoming in the Gas Hills region.  There are currently no downstream or down-gradient residences within 32 kilometers (km) (20 miles) of the facility.  The nearest residence is a ranch located approximately 6 km (4 miles) northwest of the site.  Uranium milling began at this site in 1958 and continued through 1988 with a total of 12 million tons of ore processed.  The mill utilized a conventional acid leach process.  The mill was demolished and placed in the outslope of the No. 2 Tailings Dam, with a clay-radon barrier placed over the material.  The mill area includes approximately 56 acres.  The site has three solid tailings impoundments and three tailings solution ponds.  The post-reclamation tailings piles cover approximately 241 acres.

Ground water pumping operations at the facility have been ongoing since 1980.  The corrective action consists of ground water pumping to evaporation ponds and the injection of fresh water to remove contamination and impede the flow of contaminated ground water in the aquifer.  A total of 197 million gallons of contaminated water has been collected, and 193 million gallons of fresh water injected as part of the remedial effort and approximately 217 million gallons of water have been pumped from the tailings by the end of 2001.  On December 20, 2002, ACLs were approved for the Lucky MC site, and all active correction actions ceased, such as pumping and injection.  The completion of reclamation for the one remaining settlement pond is expected in 2004.  Therefore, the license termination process is expected to begin in the fourth quarter of 2004 and should be complete in 2005.

The cost for decommissioning is estimated to be approximately $1.0 million.

## 3.0  MAJOR TECHNICAL AND REGULATORY ISSUES

None

## 4.0  ESTIMATED DATE FOR CLOSURE      2005

# PATHFINDER – SHIRLEY BASIN

## 1.0 SITE IDENTIFICATION

| | |
|---|---|
| Location: | Shirley Basin Mining District, WY |
| License No.: | SUA-442 |
| Docket No.: | 40-6622 |
| License Status: | Reclamation |
| Project Manager | Elaine Brummett |

## 2.0 SITE STATUS SUMMARY

The Shirley Basin site is located in eastern Wyoming. The former uranium mill and mine site is located approximately 5 miles northeast of the former Shirley Basin town site. The town of Shirley Basin functioned primarily as a mining camp, and has been entirely abandoned for more than 10 years. The nearest residence is a ranch located approximately three miles east of the tailings site. Uranium milling began on the site in 1971, and continued through 1992, when the last ore was processed. The site has two solid tailings impoundments, the largest covering approximately 158 acres, and the smaller 135 acres. A solution pond, which is also the disposal location for waste materials, covers about 30 acres. The mill has been decommissioned according to the DP submitted to NRC in 1992. Ground water continues to be pumped to a pond for evaporation in order to reduce contamination in the aquifer.

The cost for decommissioning is estimated to be approximately $9.0 million.

## 3.0 MAJOR TECHNICAL AND REGULATORY ISSUES

The WDEQ raised concerns about the appropriateness of the proposed ACLs for ground water in a June 5, 2003, letter to NRC after review of the Draft EA for the ACLs. The Fish and Wildlife Service and the EPA also had expressed several concerns. A partial response by the licensee was received November 14, 2003, and the final Spring Creek Assessment Plan was submitted May 26, 2004. The staff anticipates that if the creek data and WDEQ response are acceptable, the ACLs can be issued by January 2005.

## 4.0 ESTIMATED DATE FOR CLOSURE     2007

# PETROTOMICS

## 1.0 SITE IDENTIFICATION

Location:            Shirley Basin, WY
License No.:         SUA-551
Docket No.:          40-6659
License Status:      Reclamation
Project Manager:     Rick Weller

## 2.0 SITE STATUS SUMMARY

The decommissioning and reclamation of the Shirley Basin Uranium Mill, including the mill tailings impoundment, was completed in June 2001. The tailings impoundment contains 3 million tons of uranium ore tailings and covers an area of approximately 142 acres. The staff performed a final "closeout" inspection of the completed reclamation construction activities at the Shirley Basin site in July 2001. The staff completed its review of the Tailings Reclamation Construction Completion Report in February 2002 with the conclusion that reclamation of the Shirley Basin tailings was completed in accordance with the requirements of 10 CFR Part 40, Appendix A, and the licensee's approved tailings reclamation plan. The licensee is currently preparing the necessary papers to turn over ownership of the site to the DOE for long-term custody and the staff expects to terminate the Shirley Basin license in 2004.

The cost for decommissioning is estimated to be approximately $900,000.

## 3.0 MAJOR TECHNICAL AND REGULATORY ISSUES

There was one significant technical issue that was resolved at the Shirley Basin site in 2003. This issue was the State of Wyoming's concern for the potential future offsite migration of groundwater with sulphate concentrations in excess of the Wyoming standard (3,000 milligrams per liter) for Class III groundwater (livestock use). To address this concern, the licensee provided an updated sulphate transport model for the State's review in April 2003. The updated sulphate transport model indicated that sulphate concentrations in groundwater should not exceed the Wyoming Class III groundwater standard at the long-term care boundary for the site at any time in the future. The State completed its review of the updated sulphate transport model in December 2003 and determined that offsite groundwater sulphate concentrations would not exceed State standards for livestock use.

## 4.0 ESTIMATED DATE FOR CLOSURE     2004

# RIO ALGOM – AMBROSIA LAKE

## 1.0  SITE IDENTIFICATION

Location:            McKinley Co., NM
License No.:         SUA-1473
Docket No.:          40-8905
License Status:      Reclamation
Project Manager:     Jill Caverly

## 2.0  SITE STATUS SUMMARY

The site status changed from standby to reclamation in August 2003 to reflect the licensee's intent to begin full demolition and reclamation of the site leading to termination of the specific license.  The tailings impoundment contains 33 million tons of uranium ore and covers an area of approximately 370 acres.  The mill was demolished and disposed of in the tailings impoundment in late 2003.  The demolition was completed in accordance with a mill demolition plan approved by NRC in October 2003.  The staff is currently reviewing an application for ACLs.  A groundwater corrective action program is currently in effect until the review of the ACL application is completed in late 2004.  A portion of the tailings impoundment is still open for disposal of Atomic Energy Act, Section 11e.(2) byproduct material.  A final soil DP is expected to be submitted for NRC review in 2004.

The cost for decommissioning is estimated to be approximately $18 million.

## 3.0  MAJOR TECHNICAL AND REGULATORY ISSUES

None

## 4.0  ESTIMATED DATE FOR CLOSURE      2008

# SEQUOYAH FUELS CORPORATION (SFC)

## 1.0 SITE IDENTIFICATION

Location:            Gore, OK
License No.:         SUB-1010
Docket No.:          04008027
License Status:      Reclamation
Project Manager:     Myron Fliegel

## 2.0 SITE STATUS SUMMARY

Uranium and thorium contamination of the soils and subsoils has been identified at the site. In addition, the groundwater is contaminated with uranium, thorium and metals.

In March 1999, Sequoyah Fuels Corporation (SFC) submitted a DP to remediate the site and terminate the license in accordance with the 1997 LTR in 10 CFR 20.1403 for restricted conditions. In order for NRC to approve the DP, a long-term custodian to assume responsibility for institutional controls would have to be identified. However, several potentially acceptable custodians (the State of Oklahoma, the USACE, and the Cherokee Nation) declined, and SFC was unable to get a commitment from DOE.

In January 2001, SFC requested that some of the waste at the site be classified as 11e.(2) byproduct material and thus subject to Appendix A of Part 40. In July 2002, the Commission approved SFC's request that some of the wastes be classified as 11e.(2). In December 2002, the license was amended to permit possession of 11e.(2) and require site remediation in accordance with Appendix A of Part 40. After remediation and license termination, DOE will be required to assume responsibility, as the long-term custodian, if the Oklahoma chooses not to.

SFC submitted a reclamation plan in January 2003. The staff is currently reviewing the plan and developing an associated EIS. In June 2003, SFC submitted a ground water monitoring plan and a ground water corrective action plan. The staff is currently reviewing those plans.

## 3.0 MAJOR TECHNICAL AND REGULATORY ISSUES

There is significant groundwater contamination at this site which the groundwater monitoring and corrective action plan are intended to address. Staff also has concerns with the placement and design of the waste impoundment.

Financial assurance issues are summarized in SECY-03-0198 dated November 12, 2003.

A hearing has been granted to the State of Oklahoma and the Cherokee Nation on issues related to the reclamation plan proposed by SFC. Additionally, Oklahoma appealed, to the Tenth Circuit Court of Appeals, the Commission's decision regarding classification of some wastes as 11e.(2) byproduct material. Oklahoma has also petitioned for a hearing on SFC's proposed plan to dewater raffinate sludges that are currently in settlement ponds. Negotiations are being conducted by Oklahoma and SFC to resolves issues. If successful, Oklahoma could withdraw its lawsuit and hearing requests.

## 4.0 ESTIMATED DATE FOR CLOSURE      2010

**SOHIO L-BAR**

## 1.0 SITE IDENTIFICATION

Location:            Seboyeta, NM
License No.:         SUA-1472
Docket No.:          40-8904
License Status:      Reclamation
Project Manager:     Rick Weller

## 2.0 SITE STATUS SUMMARY

The decommissioning and reclamation of the L-Bar Uranium Mill, including the mill tailings impoundment, was completed in April 2000.  The tailings impoundment contains 1.7 million tons of uranium ore tailings and covers an area of approximately 100 acres.  The staff performed a final "closeout" inspection of the completed reclamation construction activities at the L-Bar site in May 2001.  The staff completed its review of the L-Bar Uranium Mill Tailings Reclamation Completion Report in September 2001 with the conclusion that reclamation of the L-Bar tailings was completed in accordance with the requirements of 10 CFR Part 40, Appendix A, and the licensee's approved L-Bar Uranium Mine Reclamation and Closure Plan.  The licensee is currently preparing the necessary papers to turn over ownership of the site to the DOE for long-term custody and the staff expects to terminate the L-Bar license in 2004.

The pending termination of the L-Bar license was delayed by the actions required by the licensee to terminate the groundwater discharge permit issued by the State of New Mexico. These actions were completed in July 2003, and the State has stated that the groundwater discharge permit will be terminated upon NRC's approval of the final LTSP for the L-Bar site which is currently being prepared by the DOE.

The cost for decommissioning is estimated to be approximately $700,000 million.

## 3.0 MAJOR TECHNICAL AND REGULATORY ISSUES

None

## 4.0 ESTIMATED DATE FOR CLOSURE     2004

# UMETCO MINERALS CORPORATION

## 1.0 SITE IDENTIFICATION

Location:          Gas Hills Region, Wyoming
License No.:       SUA-648
Docket No.:        40-0299
License Status:    Reclamation
Project Manager:   Elaine Brummett

## 2.0 SITE STATUS SUMMARY

Mill operation ended in 1984 and the mill was decommissioned in 1990. The FSSR is under NRC review, but one building in the restricted area and a small portion of the haul road have yet to be remediated. The windblown tailings remediated area is 111 acres and 4,950 cubic yards of soil were removed (soil DP approved April 2001). An additional 6,700 cubic yards of material were removed because of contamination released when the tailings dam was breached, and 30,000 cubic yards were removed from a former evaporation pond.

The covers on two disposal areas are complete, and the cover is nearly complete on the A-9 Repository (former uranium mining pit). The area of Pond 2 will be covered next year, if enough water evaporates (cover design approved November 10, 2003), and the C-18 uranium mining pit will be backfilled. The total disposal area is approximately 300 acres.

Before license termination, DOE must arrange for transfer of land that is within the long-term care boundary from the Bureau of Land Management.

The cost for decommissioning is estimated to be approximately $14.8 million.

## 3.0 MAJOR TECHNICAL AND REGULATORY ISSUES

None

## 4.0 ESTIMATED DATE FOR CLOSURE     2006

# UNITED NUCLEAR CORPORATION (UNC)

## 1.0 SITE IDENTIFICATION

| | |
|---|---|
| Location: | Churchrock, NM |
| License No.: | SUA-1475 |
| Docket No.: | 40-8907 |
| License Status: | Reclamation |
| Project Manager: | William von Till |

## 2.0 SITE STATUS SUMMARY

The facility is a conventional uranium mill site under reclamation. UNC operated the site as a uranium mill facility from 1977 to 1982. The site includes a former ore processing mill and tailings disposal area, which cover about 25 and 100 acres, respectively. The mill, designed to process 4,000 tons of ore per day, extracted uranium using conventional crushing, grinding, and acid-leach solvent extraction methods. Uranium ore processed at the site came from the Northeast Church Rock and the Old Church Rock mines. The average ore grade processed was approximately 0.12 percent uranium oxide. The milling of uranium ore produced an acidic slurry of ground waste rock and fluid (tailings) that was pumped to the tailings disposal area. Uranium milling and tailings disposal were conducted and an estimated 3.5 million tons of tailings were disposed in the tailings impoundments. The tailings disposal area is subdivided by dikes into three cells identified as the South Cell, Central Cell, and North Cell. Surface reclamation is complete, except for the area of the south tailings cell covered by two evaporation ponds, which are part of the groundwater corrective action plan.

The current effort is a groundwater corrective action plan which is also under oversight of the EPA through Superfund. A MOU has been executed between NRC and EPA for this site.

The cost for decommissioning is estimated to be approximately $3.7 million.

## 3.0 MAJOR TECHNICAL AND REGULATORY ISSUES

Upcoming groundwater corrective action issues are anticipated starting in late 2004 or early 2005. Significant coordination will be necessary with stakeholders.

The Navajo Nation has had a major interest in the site which has resulted in several NRC presentations in an effort to increase public confidence.

Sedimentation issues in the diversion channel have been communicated to the licensee as they relate to license termination.

## 4.0 ESTIMATED DATE FOR CLOSURE    2015

# WESTERN NUCLEAR, INC. – SPLIT ROCK

## 1.0 SITE IDENTIFICATION

| | |
|---|---|
| Location: | Jeffrey City, WY |
| License No.: | SUA-56 |
| Docket No.: | 40-1162 |
| License Status: | Reclamation |
| Project Manager: | William von Till |

## 2.0 SITE STATUS SUMMARY

This site is a conventional uranium mill currently under reclamation. Mill operations commenced in 1958 and continued until 1981. Uranium ore processed at the mill was extracted in mines south of the facility. The mill operations consisted of physical and chemical including sulfuric acid leaching. Decommissioning of the mill was completed on September 15, 1988. Surface reclamation is complete except for two evaporation ponds and several buildings. One large tailings pile is the main site feature covering 267 acres and weighing 12 million tons. Two evaporation ponds are currently present to support groundwater corrective action. The site has several administrative and maintenance buildings. The site is currently under groundwater remediation and a groundwater corrective action plan and application for ACLs are under review.

The cost for decommissioning is estimated to be approximately $12.3 million.

## 3.0 MAJOR TECHNICAL AND REGULATORY ISSUES

The application for a groundwater corrective action plan and ACLs was seen as a case of first impression due to the proposed use of institutional controls on off-site residential properties and an alternate water supply.

The State of Wyoming has submitted several letters outlining major concerns over the proposed groundwater corrective action plan. A meeting was held with the State in November 2003. A meeting with the licensee was held in December 2003 to discuss the proposal, staff's review status, and the licensee's progress on issues relating to an adjacent community called Red Mule. Staff will draft an EA for this action and hold a public meeting in Wyoming. Public interest has been received from the potentially affected parties.

The licensee is attempting to either acquire properties or obtain institutional controls for areas that will be impacted by site derived groundwater contamination. Staff can not complete the technical review or a draft EA until the licensee has completed this action.

## 4.0 ESTIMATED DATE FOR CLOSURE    2007

# Appendix E

# Site Summaries for
# Fuel Cycle Facilities Undergoing
# Decommissioning

# FRAMATOME RICHLAND

## 1.0  SITE IDENTIFICATION

Location:            Richland, WA
License No.:         SNM-1227
Docket No.:          70-1257
License Status:      Active
Project Manager:     Patricia Silva

## 2.0  SITE STATUS SUMMARY

This facility has five lagoons which were used as part of the waste-water treatment process. The Washington State Department of Ecology ("Ecology") has ordered the licensee to close the lagoons.  The licensee and Ecology entered into a consent decree agreement identifying enforceable milestones for completion of closure of the lagoons.  These milestones include emptying the lagoons by September 8, 2004 and submitting a closure certificate for the lagoons by August 8, 2006.

On March 31, 2004, an amendment to the Framatome license was granted which would phase out requirements to conduct monthly between-liner sampling as lagoons were emptied of their inventory.  This sampling activity monitored the effectiveness of the lagoon upper liners in containing the lagoon inventories without significant leakage to the interstitial space between the upper and lower liners.

As of 2004, the lagoons have been removed from service and have been sealed off from process streams.  All removable liquid has been removed and processed.  The licensee continues to address rainwater accumulation as needed.  Sludge has been removed from 3 out of 5 lagoons.  Removal of all lagoon inventories is scheduled for completion by the agreed date of September 8, 2004.

Activities planned from 2004 until the closure deadline of August 8, 2006 include characterization, remediation, and removal of soil between the liners, and soil below the liners, as needed.

## 3.0  MAJOR TECHNICAL OR REGULATORY ISSUES

Framatome has indicated that they will be requesting an amendment to their special nuclear material (SNM) license to remove the enrichment check in the solids processing facility tank for the lagoon soils wash process in order to meet deadlines identified in the Ecology consent decree agreement.  Because of mandated completion dates for various phases of this activity, Framatome will be asking for an amendment within 60 to 90 days after submittal of the request.

## 4.0  ESTIMATED DATE FOR CLOSURE

Framatome has a binding agreement with Ecology to submit a closure certificate by August 8, 2006.

# GENERAL ATOMICS

## 1.0  SITE IDENTIFICATION

| | |
|---|---|
| Location: | San Diego, CA |
| License No.: | SNM-696 |
| Docket No.: | 70-734 |
| License Status: | Active |
| Project Manager: | Julie Olivier |

## 2.0  SITE STATUS SUMMARY

In September 1996 General Atomic's (GA's) SNM license was amended to authorize only decommissioning activities.  By an application dated October 11, 1996, and supplements dated December 5, 1996; April 18, 1997; and January 15, 1998; GA requested an amendment to its license to incorporate a site DP.  In accordance with the DP GA is undergoing site wide decommissioning.  Six areas were released in 2004.  There are two more areas to be decommissioned and released between 2004 and 2005.  GA plans to submit a license request to lower the possession limit to less than a critical mass in 2004, after which time the site will be transferred to DWMEP.

The primary radioactive contaminant is uranium-235.  Soil will be remediated to levels specified in option 1 of the BTP, "Disposal or Onsite Storage of Thorium or Uranium Wastes from Past Operations," [46 FR 52061; October 23, 1981].  Facilities and equipment will be decontaminated to levels specified in "Guidelines for Decontamination of Facilities and Equipment Prior to Release for Unrestricted Use or Termination of Licenses for Byproduct, Source, or Special Nuclear Material," [USNRC, Policy and Guidance Directive FC 83-23, Division of Industrial and Medical Nuclear Safety, November 4, 1983].  GA intends to decommission to radiation levels required for unrestricted use and to terminate its SNM license.

## 3.0  MAJOR TECHNICAL OR REGULATORY ISSUES

None

## 4.0  ESTIMATED DATE FOR CLOSURE      Termination of Part 70 license is 2006.

# HONEYWELL

## 1.0 SITE IDENTIFICATION

Location:            Metropolis, IL
License No.:         SMB-526
Docket No.:          40-3392
License Status:      Active
Project Manager:     John Lusher

## 2.0 SITE STATUS SUMMARY

This facility is the only operational conversion facility in the United States. There are two $CaF_2$ settling ponds on this site. In CY 2001, NRC determined that the material in the ponds could be treated as exempt material, as defined in 10 CFR 40.13(a), and should be disposed of accordingly.

In CY 2003, the licensee remediated "A" pond and disposed of the solid material. Treatable liquid has been removed from "C" pond, but the licensee is planning to treat the solids differently than in the past. The licensee is planning a significant engineering project to replace its current effluent treatment facility with a system capable of recovering uranium and calcium solids that will not utilize a lagoon component.

## 3.0 MAJOR TECHNICAL OR REGULATORY ISSUES

None

## 4.0 ESTIMATED DATE FOR CLOSURE    2020

NRC FORM 335
(9-2004)
NRCMD 3.7

U.S. NUCLEAR REGULATORY COMMISSION

**1. REPORT NUMBER**
(Assigned by NRC, Add Vol., Supp., Rev., and Addendum Numbers, if any.)

NUREG-1814

# BIBLIOGRAPHIC DATA SHEET

*(See instructions on the reverse)*

**2. TITLE AND SUBTITLE**

Status of Decommissioning Program: 2004 Annual Report

| 3. DATE REPORT PUBLISHED | |
|---|---|
| MONTH | YEAR |
| January | 2005 |

**4. FIN OR GRANT NUMBER**

**5. AUTHOR(S)**

J. Buckley

**6. TYPE OF REPORT**

FINAL

**7. PERIOD COVERED** *(Inclusive Dates)*

**8. PERFORMING ORGANIZATION - NAME AND ADDRESS** *(If NRC, provide Division, Office or Region, U.S. Nuclear Regulatory Commission, and mailing address; if contractor, provide name and mailing address.)*

Division of Waste Management and Environmental Protection

Office of Nuclear Material Safety and Safeguards

Nuclear Regulatory Commission

Washington, DC 20555-0001

**9. SPONSORING ORGANIZATION - NAME AND ADDRESS** *(If NRC, type "Same as above"; if contractor, provide NRC Division, Office or Region, U.S. Nuclear Regulatory Commission, and mailing address.)*

Same as above

**10. SUPPLEMENTARY NOTES**

John T. Buckley, NRC Project Manager

**11. ABSTRACT** *(200 words or less)*

This report provides a comprehensive overview of the U.S. Nuclear Regulatory Commission's (NRC's) decommissioning program. Its purpose is to provide a stand-alone reference document which describes the decommissioning process and summarizes the current status of all decommissioning activities including the decommissioning of complex decommissioning sites, commercial reactors, research and test reactors, uranium mill tailings facilities, and fuel cycle facilities. In addition, this report discusses accomplishments in the decommissioning program since publication of the 2003 annual report (SECY-03-0161), and it identifies the key decommissioning program issues which the staff will address in fiscal year (FY) 2005.

**12. KEY WORDS/DESCRIPTORS** *(List words or phrases that will assist researchers in locating the report.)*

decommissioning, 2004 annual report, nuclear power reactors, research and test reactors, complex decommissioning materials facilities, uranium recovery facilities, fuel cycle facilities

**13. AVAILABILITY STATEMENT**

unlimited

**14. SECURITY CLASSIFICATION**

*(This Page)*

unclassified

*(This Report)*

unclassified

**15. NUMBER OF PAGES**

177

**16. PRICE**